T0236510

SpringerBriefs in Applied Sciences and Technology

More information about this series at http://www.springer.com/series/8884

Soumya Sen · Agostino Cortesi · Nabendu Chaki

Hyper-lattice Algebraic Model for Data Warehousing

 Springer

Soumya Sen
University of Calcutta
Calcutta, West Bengal
India

Nabendu Chaki
University of Calcutta
Calcutta, West Bengal
India

Agostino Cortesi
Università Ca' Foscari Venezia
Venice
Italy

ISSN 2191-530X   ISSN 2191-5318   (electronic)
SpringerBriefs in Applied Sciences and Technology
ISBN 978-3-319-28042-4   ISBN 978-3-319-28044-8   (eBook)
DOI 10.1007/978-3-319-28044-8

Library of Congress Control Number: 2015958340

Printed on acid-free paper

This Springer imprint is published by SpringerNature
The registered company is Springer International Publishing AG Switzerland

"Dedicate this work as a tribute to the immortal soul of my parents (Late Rajendra Narayan Sen and Late Sarmila Sen). Their blessings were, as always, the prime mover towards attainment of my goal."

Soumya Sen

Preface

It is an immense pleasure to introduce this book on data warehousing that emphasizes on new algebraic structure and on using appropriate query language for efficient extraction of information from a multi-level storage abstraction. This book is a handbook for future researchers who want to work on data warehousing from modeling perspective.

This book focuses on two inherent structures of data warehousing, namely lattice of cuboids and concept hierarchy. We analyze the lattice as an algebraic structure and identify its static nature as a problem. We discuss this problem in the domain of data warehouse. However, the limitation of lattice due to its static nature also persists in other application domains. This problem identified in this book is resolved by a novel algebraic model named *Hyper-lattice* introduced in this text. We define the properties, propositions of this new structure, and also proof the lemmas on it. We theoretically establish *Hyper-lattice* as a standard algebraic model and explain its flexibility over the lattice.

As this book is on data warehousing, we show the formation of the algebraic structure as a *Hyper-lattice of cuboids*. Different case studies illustrate and establish the novelties and flexibility of Hyper-lattice of cuboids over the traditional lattice of cuboids.

Furthermore the analysis is being continued in this book on concept hierarchy to manage the dimension from different abstract levels. Concept hierarchy provides the capability to view the same data from various abstractions and is represented as a hierarchy. We take this opportunity to traverse the hierarchical structure both in top-down and in bottom-up manner as required to solve a problem. We have used cooperative query language to express the result of fetching the data from concept hierarchy. Here, we show the advantage of using cooperative query language over traditional SQL and also explore how to use multiple abstractions of a dimension in a single query.

The book is organized into four chapters. In the first chapter, we propose and establish Hyper-lattice as an algebraic model and show its effectiveness in organizing the cuboids data warehousing. In the second chapter, real-life case studies are described to illustrate the use of Hyper-lattice in data warehousing application.

In the third chapter, we discuss and propose a framework to traverse the concept hierarchy efficiently and use cooperative query language to generate useful and more information. Finally, we conclude our findings in the fourth chapter.

We thank and appreciate Prasun Naiya, Tanusri Kundu, and Jayshree Nath for their contributions in implementing the proposed methodologies. Our special thanks to Mary James and her colleagues in Springer for being so patient and cooperative during preparation of the chapters. Without this care and support, this book was never possible. Last but not the least, we thank our family members for their valuable support to concentrate on this book. We expect our book would be fruitful for the budding research scholars who want to explore the area of data warehousing and OLAP.

Soumya Sen
Agostino Cortesi
Nabendu Chaki

Contents

Chapter 1
Hyper-lattice

1.1 Introduction

In this chapter we introduce and discuss the notion of hyperlattice, a generalization of the notion of lattice of cuboids which is at the basis of most data warehousing techniques.

Lattice of cuboids is an important structure in data warehouse. The number of cuboids in this structure grows exponentially with addition of newer dimensions. N-dimensional lattice of cuboids consists of 2^N numbers of cuboids. In Fig. 1.1, we see a lattice of cuboids with 3 dimensions namely A, B and C. It consists of 8 cuboids as the number of dimension is 3. Here, <A,B,C> is the base cuboid and <All> is the apex cuboid for the lattice.

Managing this large structure in terms of different operations ensuring efficient access mechanism is a subject of interest. In this context, we introduce a mechanism of identifying the dimension and cuboid in terms of a numeric value called Tag Number [1] and Tagged Value [1].

Tag Number is used to uniquely identify a dimension of a cuboid in the lattice using an integer value. The first dimension has tag number 0 and for every new dimension tag number is increased by one.

Tagged Value is used to uniquely identify individual cuboids in a lattice. This is computed by taking the tag numbers as exponents in a sum of power of 2.

In Fig. 1.1 consider A, B and C has Tag number 0, 1 and 2. Hence the tagged value of cuboid <A,B,C> is $2^0 + 2^1 + 2^2 = 7$. Similarly the tagged values for <A,B>, <A,C>, <B,C>, <A>, <B>, <C> are 3, 5, 6, 1, 2 and 4 respectively.

Besides, one can view a lattice of cuboids as a hierarchical graph with multiple paths to reach a cuboid from another cuboid. These alternative paths are the different options of summarization from a lower to an upper level cuboid. The alternate paths may involve different amounts of storage space and different volume of computations. Thus the finding the minimal path is important.

© The Author(s) 2016
S. Sen et al., *Hyper-lattice Algebraic Model for Data Warehousing*,
SpringerBriefs in Applied Sciences and Technology,
DOI 10.1007/978-3-319-28044-8_1

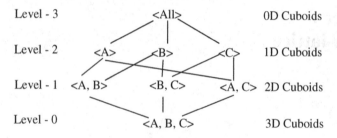

Fig. 1.1 Lattice of cuboids with dimensions A, B and C

1.2 Related Study

Optimization on the lattice of cuboids has been studied for a long time in the data warehouse domain from different perspectives.

In order to improve storage and retrieval of data cubes, the lattice of cuboids are explored since the inception of the OLAP system. There have been a number of works on computational aspects of cube, aggregation, multi-dimensionality and also on the OLAP operations such as roll-up, drill-down etc. [2–10]. A straightforward approach (named Independent) is presented in [2] to independently compute each cuboid from the base cuboid, using standard group-by techniques (hash based or sort based). Thus the base cuboid is read and processed for each cuboid to be computed. This is a static method and this method leads to poor performance both in terms of space and time complexity. Another optimization method (named Amortize-scans) presented in [2] aims at amortizing disk reads by computing as many group-bys as possible, together in memory. Let us consider a cuboid ABCD. If the group-by of ABCD is stored on disk, to reduce disk read costs of the cuboids ABC, ACD, ABD and BCD are to be computed in one scan of ABCD. Share-sorts optimization technique [2] is specific to the sort-based algorithms and aims at sharing sorting cost across multiple group-bys. When a hash-table is too large to fit in memory, data is partitioned. Subsequent aggregation is done for each partition that fits in memory. This kind of share-partitioning optimization [2] is specific to the hash-based algorithms.

Time-series data [4] often defines the hierarchy of multi-dimensional data. However, the hierarchy and dependencies between the cuboids is not necessarily on time series alone [5, 7, 10]. There has been work on the efficiency of OLAP operations on spatial data in a data warehouse [8]. The efficiency of roll-up and drill-down on the data collection stored in the relational database management system has been considered [9]. A method using labeling on a tree structure has been proposed [19] and found to be advantageous over more conventional mapping tables to store the hierarchy.

A lattice framework has been proposed to express dependencies [11] among views. Greedy algorithms that work off this lattice and determine a good set of views to materialize is presented. The greedy algorithm performs within a small constant factor of optimal criterion under a variety of models. The most common of the hypercube lattice is considered and the choice of materialized views for hypercube in detail is examined, giving some good tradeoffs between the space used and the average time to answer a query. The technique proposed in [12] reduces the solution space by considering only the relevant elements of the multidimensional lattice. An additional statistical analysis allows a further reduction of the solution space. A greedy approach is presented [13] that turns out to be very effective; it is both polynomial-time as a function of the number of possible views to materialize and guaranteed to come close to the optimum choice of views. Parallel computation is another approach that has been applied for optimization of cuboids. A multi-cube computation [14] approach proposes the concept of processing multiple cubes simultaneously. This methodology used both sort and hash based approach. Another method of parallelism [15] using bottom-up and top-down method was proposed and implemented in shared disk type parallel machines consisting of numbers of processors. In another research work named as cgmCUBE [16] parallel data cube generation was used to optimize the ROLAP. Here computation is organized in task graph and then graph partitioning is used to partition the task to each node.

Some of the research works focus on modification of lattice structure for efficient computation such as iceberg cube [17]. Two of the major practices in this context are top-down computation and bottom-up computation. Top-down computation is represented by multi-way array aggregation [18] which is based on shared computation [19]. Bottom-up computation is represented by BUC [20], which is actually based on a priori pruning [21]. Multi-way array aggregation performs well on dense data sets whereas BUC is suitable for sparse data set. Combining these two approaches Star-Cubing [22] was proposed, even though its performance is not satisfactory for very sparse data due to the maintenance of tree structure. An alternative approach, MM-Cubing [23], works fine with different types of distribution of data: it computes Iceberg cubes by factorizing the lattice space according to the frequency of values. A new version of Star-Cubing [24] explores multi-dimensional simultaneous aggregation using star-tree methods. The performance analysis shows the superiority of this method over the existing methods.

Along with these studies, some of other methods also require special mention those are based on different mathematical approaches. A log-linear model [25] of compressing data cube is one of them. Another approach uses the concept of data cube approximation [26] via wavelets.

Data modelling is another important consideration in lattice management. In this context it is studied under dimension modelling [27], multi-dimensional analysis [28]. The lattice of cuboid is originated from base cuboid, which corresponds to a fact table and the fact table itself is associated with related dimension tables. In different applications of data warehouses different models are proposed to work for that particular application. Whenever any data warehouse schema or model is considered and any modification or enhancement is done it directly or indirectly refers to the lattice structure. However majority of the research works in this area are concerned about the particular applications only, not modelling the lattice in general. Some of them include weather forecasting [27], medical waste management [28] etc. Spatial data is also considered for building warehouses and researches show inclusion of spatial dimension enable spatial analysis. A multi dimensional design framework is presented [29]. It is adapted for effective spatial-temporal exploration and analysis through the extension of a conceptual model with spatial dimensions to enable spatial analysis.

The above survey works present different techniques of optimization of data warehouse cubes as well other efficient methods of representing the lattice of cuboids.

1.3 Problems of the Existing Structure

Data warehouse schemas are being built based on the requirement of organization. In data warehouse schema fact table is being built with the associated dimension tables. The fact table corresponds to the base cuboid in the lattice of cuboids. The entire lattice of cuboids is being formed the base cuboid by applying series of roll-up operations. Hence the entire structure is dependent on the data warehouse schema. This suggests that the construction of lattice of cuboids is being "frizzed" once the data warehouse schema is being finalized. However in the real life, the business applications change frequently. The changes may force to introduce new dimensions for the business analysis. This problem leads to two solutions. First, addition of a new dimension can be performed at the base cuboid of the existing lattice, which simply doubles the structure.

Unfortunately, all the new cuboids that are generated may not be subject of interest for business analysis. Still they are being created and stored in the system. Second approach is required, if the addition of new dimension is relevant to some of the dimension of existing lattice of cuboids. In this case a separate lattice of cuboids is created based on the new dimension and some of the existing/ old dimensions. The disadvantage of the first solution is the creation of a numbers of additional cuboids and the disadvantage of the second solution is creation of multiple lattice of cuboids where some of the cuboids are overlapping. The above facts identify the limitations of lattice in representing the dynamic insertion of new dimension to support the changing the business requirements. Lattice, as an algebraic structure suffers from scalability, and lack of flexibility in such cases.

On the fly analytic decision making, finding a better information retrieval strategy depending on the current context, and scalability of solutions are a few desirable features for any computer based mathematical model. This has been the primary motivation towards finding a more suitable algebraic structure that would replace or modify the existing lattice structure to adapt the dynamics of the business data being analyzed.

1.4   A Brief Idea Towards the New Structure

In order to increase flexibility of analytical processing, it's required to allow insertion of new dimension(s) at any existing cuboid except the apex and the base
cuboids. Whenever the new dimension is added to a particular cuboid, the dimension of that cuboid is increased by 1. Hence, this new cuboid is placed one level
below the target cuboid. As this cuboid contains a new dimension, the cuboids that
could be generated from it through roll-up would also contain some new cuboids
which are not part of the existing structure. The newly created cuboid is at the
level below of target cuboid and this act as the base cuboid for all the cuboids in a
hierarchy. This structure is no longer a lattice.

Let's consider a lattice with 3 dimensions A, B and C. Say a new dimension D
is added at cuboid <B,C> at level-1. This would be considered in this structure as
a new base cuboid at level-0 as <B,C,D>. Figure 1.2 depicts the modified lattice
structure with new base cuboid <B,C,D>.

Starting from this base cuboid <B,C,D>, 3 cuboids are generated through roll-
up operation at level-1 namely <B,C>, <B,D> and <C,D>. The cuboid <B,C>
is already there level-1. The two new cuboids <B,D>, and <C,D> would also
be placed at level-1. Again roll-up operations would be applied on the cuboids
<B,C>, <B,D> and <C,D> to generate 1-dimensional cuboids at level-2. Only the
cuboid <D> would be added at level-2 as the two other 1-dimensional cuboids
<B> and <C> are already in level-2. The dimension D from level-2 would also be
connected to the apex cuboid at level-3. Thus we find that the resultant structure
as shown in Fig. 1.3. Interestingly, a new lattice has formed by the elements in the
ascending chains linking <B,C,D> to <All>.

Hence the target of these series of roll-up operations is to generate the missing
cuboids and finally this concept leads us to conceptualize a model which is a collection of overlapping lattices. In Fig. 1.3 we depict a pictorial view of the new
concept.

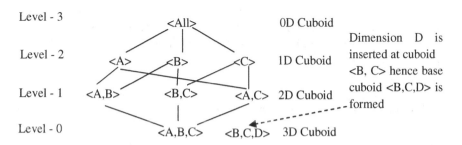

Fig. 1.2   3 dimensional lattice <A,B,C> with another base cuboid <B,C,D>

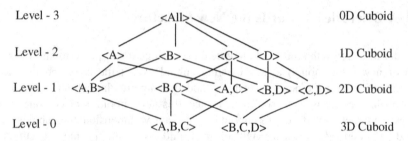

Fig. 1.3 Overlapping of two sets of lattice of cuboids

1.5 Introduction of Hyper-lattice

The idea that has been floated in the previous sub-section motivates us to define, formulate and proof its algebraic properties to establish it as a mathematical model which is more flexible than the traditional lattice. This new algebraic structure is proposed as Hyper-lattice. In the following paragraphs, we have formally defined Hyper Lattice and the concept behind.

Definition of Hyper-lattice

A partially ordered set A is a Hyper-lattice if: (i) every subset of A has a unique least upper bound (LUB), and (ii) there is a subset L of minimal elements in A such that for each c in L, the set of elements in the ascending chains from c to LUB(A) is a lattice.

The interesting aspect of this newly proposed algebraic structure is that the Hyper-lattice in Fig. 1.3 is actually the union of two lattices L1 and L2 as listed below with the same LUB of <All>.

L1: {<A,B,C>, <A,B>, <A,C>, <B,C>, <A>, <B>, <C>, <All>}
L2: {<B,C,D>, <B,C>, <B,D>, <C,D>, <B>, <C>, <D>, <All>}

Another important observation is that any new dimension may be added more than once in the Hyper-lattice. If dimension D is further added in the cuboid <A>, then another new base cuboid <A,D> would be added in the Hyper-lattice. This is depicted in Fig. 1.4.

1.5.1 Conceptual View of Hyper-lattice

A Hyper-lattice consists of at least two overlapping lattices where some of the elements are common and must meet at common point which is the LUB of each of the lattice however the GLB of each lattice should be unique.

We conceptualize this idea of Hyper-lattice in terms of a diagram in Fig. 1.5. Initially we have a lattice A. It is merged with another lattice B which has a

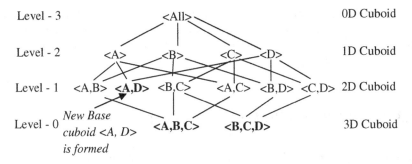

Level - 3 ⟨All⟩ 0D Cuboid

Level - 2 ⟨A⟩ ⟨B⟩ ⟨C⟩ ⟨D⟩ 1D Cuboid

Level - 1 ⟨A,B⟩ ⟨A,D⟩ ⟨B,C⟩ ⟨A,C⟩ ⟨B,D⟩ ⟨C,D⟩ 2D Cuboid

Level - 0 *New Base cuboid ⟨A, D⟩ is formed* ⟨A,B,C⟩ ⟨B,C,D⟩ 3D Cuboid

Fig. 1.4 Hyper-lattice with 3 base cuboids ⟨A,B,C⟩, ⟨B,C,D⟩ and ⟨A,D⟩

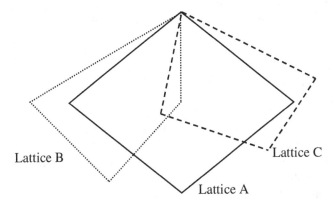

Lattice B

Lattice C

Lattice A

Fig. 1.5 A hyper-lattice of 3 lattices A, B and C

different GLB of A, but has few elements in common. This forms a Hyper-lattice. Thereafter it is merged with another lattice C, which has GLB other than of A and B but has common elements of the existing Hyper-lattice consisting of A and B. Thus we have a Hyper-lattice consisting of 3 lattices namely A, B and C.

1.5.2 Use of Hyper-lattice to Reduce Time and Space Complexity

Use of Hyper-lattice is going to save both space and time. If we refer to the Fig. 1.5 we can find that the different lattices share a common space. Hence the elements which reside in the common space need to be stored only once instead of storing separately for individual lattices. This saves space and if we think of deploying this structure from data storage perspective it helps to avoid redundancy and therefore ensures the consistency of the data. Consequently, in terms of access time, Hyper-lattice saves us from searching the elements from different lattices stored in separate memory locations and are not anyway related. Using Hyper-lattice, all the lattice of cuboids are stored within the same storage structure which makes it easy and fast for searching the data.

1.5.3 Hyper-lattice to Replace Multiple Lattices

Hyper-lattice is established as collection of multiple lattices which shares some common elements. This gives us an opportunity to merge the existing lattices. Therefore in a data warehouse application if multiple lattice of cuboids exist separately, now we can merge them if they have some common dimensions.

Say two separate lattice of cuboids exist in a system. The base cuboid of one lattice is <A,B,C,D> and of another <C,D,E>. For the first lattice of cuboids the set of cuboids are {<A,B,C,D>, <A,B,C>, <A,B,D>, <A,C,D>, <B,C,D>, <A,B>, <A,C>, <A,D>, <B,C>, <B,D>, <C,D>, <A>, <B>, <C>,<D>} and for the second lattice of cuboids the cuboids are {<C,D,E>, <C,D>, <C,E>, <D,E>, <C>, <D>, <E>}.

If we merge these two lattices we would get a Hyper-lattice where the cuboids would be stored across level-0 to level-4. One base cuboid <A,B,C,D> would be at level-0 and another base cuboid <C,D,E> would be at level-1. We find that these two set of cuboids have <C,D>, <C> and <D> in common. Therefore instead of storing these 3 cuboids twice we can store them only once.

As mentioned in Sect. 1.5.2, the Hyper-lattice structure would be cost effective in terms of both space and time.

1.6 Creation and Maintenance of Hyper-lattice

The proposed algebraic structure is conceptualized here as a collection of lattices. However, apparently this causes degeneration of the lattice structure. The collection of lattices is termed as Hyper-lattice in earlier this section. Initially it starts with a single lattice. The base cuboid of this lattice acts as parent base cuboid. If at the beginning there are N dimensions it would form a lattice of 2^N cuboids. Later on due to business requirements new dimensions may appear in the system. The challenge here is how to represent these new dimensions in the algebraic structure. A new dimension could be added here for the analysis purpose between the level 1 to $(N - 1)$. Whenever a new dimension comes to the system it is required to specify along with which existing cuboid (target cuboid) the analysis would proceed further. However, a new dimension may be inserted in more than one places of hyper-lattice structure. In this context, we introduce two new tables namely Tag Number Table (TNT) and Base Cuboid Table (BCT).

(A) **Tag Number Table (TNT)**: It is a two column table. The 1st column stores the name of the dimensions of lattice and Hyper-lattice and the 2nd column stores the Tag number associated with it.

(B) **Base Cuboid Table (BCT)**: In a Hyper-lattice there is more than one base cuboid. All these base cuboids are stored in Base Cuboid Table. This is a three column table. The 1st column stores the name of the base cuboid (or the pattern of the base cuboid), 2nd column stores the Tagged value for it and the 3rd column stores the level of the base cuboid.

If a new dimension is to be added, the name of this new dimension and its tag number is inserted in the Tag Number Table (TNT). If the tag number of the last row of TNT is t, the tag number of the newly inserted dimension is $(t + 1)$. This newly inserted dimension would be placed in the existing lattice of cuboids, as a new cuboid. This new cuboid is formed by adding this new dimension with the target cuboid at a level below the point of insertion of the target cuboid. This means if the target cuboid is at level m the newly formed cuboid would be added at level $(m - 1)$. This new cuboid would be a new base cuboid for an overlapping lattice in the Hyper-lattice. The tagged value of this new cuboid is computed, and information about this newly formed base cuboid is stored in the Base Cuboid Table (BCT). Now, by applying the roll-up operations on the new base cuboid, a new lattice is formed. The cuboids which are already present in the existing structure are not generated twice. The newly formed lattice along with the existing lattice would form a Hyper-lattice. The same steps would be repeated for adding every new dimension. The Hyper-lattice will have more overlapping lattices.

This system has an assumption that no new dimension could be added to the cuboids at level 0 or level N. This assumption is required for two reasons. If a dimension is added at level-0 then the new base cuboid would be created at (level-1) (negative level value) which is impractical in this context. Moreover this structure is originated from parent base cuboid which is at level 0. Any level below this would totally break this structure. The highest allowable level is $(N - 1)$, because the only level above it only contains apex cuboid which is a measure value only. Thus adding a dimension to apex cuboid makes no sense here. Due to this reason highest allowable level value is $(N - 1)$.

It's to be noted here, that a base cuboid which has been added except at level-0 may not remain a base cuboid later, if a dimension is added to that existing base cuboid and therefore a new base cuboid would be formed. The old base cuboid would be present as other cuboids in the newly formed Hyper-lattice. Hence the old base cuboid would be deleted from Base Cuboid Table (BCT) and the new one would be added.

1.6.1 Proposed Algorithm Create Hyper-lattice

Repeat the following steps for every DIM (*DIM represents Dimension*)

Step 1: IF DIM comes to the system for the first time then computes the Tag Number of DIM. This is calculated by increasing the value of last tag number stored in TNT by 1. The name of DIM and its Tag Number is stored in TNT;

Step 2: A new cuboid say NEWCUB is formed by combining the target cuboid TAR at level T and the new dimension DIM;

Step 3: NEWCUB is placed at level $(T - 1)$;

Step 4: NEWCUB would act as a base cuboid of an overlapping lattice in the Hyper-lattice;

Step 5: Tagged Value of NEWCUB, say TG, is computed;

Step 6: Name of NEWCUB, its level in hyper-lattice that is $(T - 1)$ and the Tagged Value TG are stored in Base Cuboid Table (BCT);

Step 7: IF TAR is already a base cuboid then delete the entry of TAR from BCT./*As TAR does not remain a base cuboid*/

Step 8: Apply roll-up operations on base cuboid NEWCUB to generate all the cuboids to form the lattice of cuboids based on NEWCUB. However the cuboids that are already present (overlapping cuboids) in the Hyper-lattice they are not re-generated;

End of Repeat

1.6.2 Illustrative Example

Case 1

Initially there is a lattice of 4 dimensions namely A, B, C and D. The base cuboid of this lattice is <A,B,C,D>. Tag number of A, B, C and D are 0, 1, 2 and 3 respectively. The tagged value of base cuboid is computed as $2^0 + 2^1 + 2^2 + 2^3 = 15$. As the base cuboid <A,B,C,D> is the base cuboid of the original lattice this one is parent base cuboid. At this stage Tag Number Table (TNT) and Base Cuboid Table (BCT) would be as follows (Tables 1.1, 1.2, 1.3, and 1.4) (Fig. 1.6).

Applying roll-up operation on newly formed base cuboid <A,B,E>, new dimension pattern <A,B>, <A,E> and <B,E> are generated at level 2. Cuboid <A,B>

Table 1.1 Tag number Table-I

Dimension name	Tag number
A	0
B	1
C	2
D	3

Table 1.2 Base cuboid
Table-I

Base cuboid pattern	Tagged value	Level of the base cuboid
<A,B,C,D>	15	0

Table 1.3 Modified tag
number Table-I

Dimension name	Tag number
A	0
B	1
C	2
D	3
E	4

Table 1.4 Modified base
cuboid Table-I

Base cuboid pattern	Tagged value	Level of the base cuboid
<A,B,C,D>	15	0
<A,B,E>	19	1

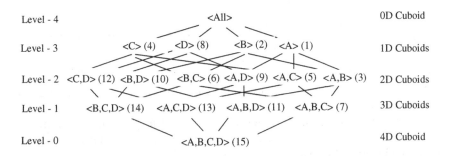

Fig. 1.6 Lattice of cuboids with tagged value

is already present and thus <A,E> and <B,E> are to be added in the immediate
upper level. Subsequently the newly generated dimension patterns are added at the
higher level of Hyper-lattice.

The hyper-lattice would look as shown in Fig. 1.7. This Hyper-lattice consists
of two overlapping lattices with the two based cuboids <A,B,C,D> and <A,B,E>.
These have been marked in two different outlines. Similarly, even newer dimensions may still appear in Fig. 1.7 in levels 1 through 3.

Case 2

Here we start with a 3 dimensional lattice A, B and C. Tag Number of A, B and
C is 0, 1 and 2 respectively. The tagged value of the base cuboid <A,B,C> is 7.
The base cuboid <A,B,C> would be parent base cuboid of the structure. At this
stage Tag Number Table (TNT) and Base Cuboid Table (BCT) would be as follows (Tables 1.5 and 1.6).

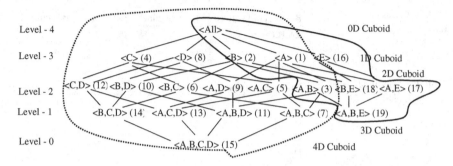

Fig. 1.7 Hyper-lattice of cuboids with tagged value

Table 1.5 Modified base
cuboid Table-I

Dimension name	Tag number
A	0
B	1
C	2

Table 1.6 Base cuboid
Table-II

Base cuboid pattern	Tagged value	Level of the base cuboid
<A,B,C>	7	0

Now a new dimension D is added to the cuboid A of the lattice. This is going
to create a new base cuboid <A,D> at level-1. Applying the roll-up operation
on <A,D> a new cuboid <D> would be generated at level-2. The newly created
Hyper-lattice is shown in Fig. 1.8. The modified Tag Number Table (TNT) and
Base Cuboid Table (BCT) are given Tables 1.7 and 1.8.

Later the dimension B is further added to the cuboid <A,D>. This is going to
create a new base cuboid <A,B,D> at level-0 and would be inserted in BCT.
Moreover as this insertion is taken place in an existing base cuboid <A,D>, this one
would be deleted from Base Cuboid Table (BCT). No further insertion would take
place in Tag Number Table (TNT), as B is already present in the system. The new
Hyper-lattice along with new BCT is shown in Fig. 1.9 and Table 1.9 respectively.

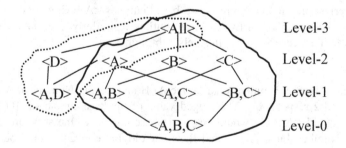

Fig. 1.8 Hyper-lattice of cuboids

Table 1.7 Modified tag number table for the hyper-lattice of Fig. 1.8

Dimension name	Tag number
A	0
B	1
C	2
D	3

Table 1.8 Modified base cuboid table for the hyper-lattice of Fig. 1.8

Base cuboid pattern	Tagged value	Level of the base cuboid
<A,B,C>	7	0
<A,D>	9	1

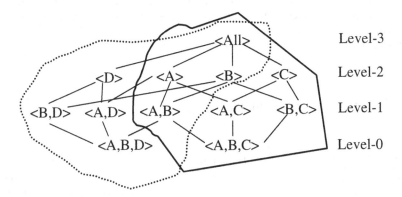

Fig. 1.9 Updated hyper-lattice of Fig. 1.8

Table 1.9 Modified base cuboid table corresponds to the hyper-lattice of Fig. 1.8

Base cuboid pattern	Tagged value	Level of the base cuboid
<A,B,C>	7	0
<A,B,D>	11	0

1.7 Propositions

1. A Hyper-lattice A having a (unique) least element is a lattice.
2. Let S be a partial order, and let B be a subset of S. The (upper-)closure of B with respect to set union is a Hyper-lattice.
3. Let A be a Hyper-lattice, and let b be a minimal element of A. Then A-{b} is still a Hyper-lattice.
4. Let S be a partial order, and let A be a Hyper-lattice subset of S. If there exists an element s in S and a minimal element b of A such that s < b, then A ∪ {s} is a Hyper-lattice too.

1.8 Other Important Properties

In this section, a few more properties of the proposed algebraic structure are presented. These are trivial for a formal proof and listed below.

1. A Hyper-lattice consists of one or more constituent lattices.
2. Any numbers of dimensions could be added to the Hyper-lattice.
3. New dimension could be added between level 1 to level $(N - 1)$ of Hyper-lattice.
4. Numbers of base cuboids would increase with the addition of each new dimension in the structure, while there would always be a single apex cuboid for the entire Hyper-lattice.
5. The number of levels for a Hyper-lattice is constant and it's the same as the number of dimensions in the base cuboid with maximum number of dimensions.
6. A new dimension could be added more than once in the Hyper-lattice structure.

1.9 Hyper-lattice Schema

An interesting feature that we can identify from Hyper-lattice is that as the structure consists of multiple fact tables a new complex schema structure is present. However each fact table is in the form of either in star schema or snowflake schema. All of these schemas are present in a single structure and we define this conglomerated structure as Hyper-lattice schema.

A Hyper-lattice schema is composition of existing data warehouse schemas where *more than one fact tables* are present as different base cuboids at same or different levels of the Hyper-lattice. These are related with each other by sharing dimensions.

Given a set of dimensions Dim $= \{D_i: i \in I\}$, and a fact table T, we denote by *dimensions*(T) the subset of Dim of dimensions represented in T

Now we formally define the Hyper-lattice schema.

Hyper-lattice schema: Given a set of dimensions Dim $= \{D_i: i \in I\}$, and a set of fact tables $\{T_j: j \in J\}$ on dimensions in Dim such that for each pair of indexes h, k, if h $\neq$ k then *dimensions*(T_h) $\neq$ *dimensions*(T_k), for each set of dimension $S \subseteq \cup\{dimensions(T_j): j \in J\}$ consider the table R_S that gives access to the tuples in the fact tables corresponding to values in dimensions in S. The Hyper-lattice schema on $\{T_j: j \in J\}$ is a directed graph having as nodes the elements in set $\{T_j: j \in J\} \cup \{R_S: S \subseteq \cup\{dimensions(T_j): j \in J\}\}$, and as edges the pairs (R_S, T_j) such that $S \subseteq dimensions(T_j)$.

1.9.1 Properties of Hyper-lattice Schema

1. Each fact table is differentiated from other fact tables, at least by one new dimension Table
2. The number of associated dimension tables with each fact table can be different in number
3. Minimum numbers of dimensions associated with a fact table is 2 and maximum is N, where N is numbers of levels in the Hyper-lattice.
4. The base cuboid of the fact tables exist at different level of Hyper-lattice.
5. A fact table of Hyper-lattice schema may not remain a fact table later, if a dimension is added to it. However, if a fact table is at level-0, it would remain as fact table always.

1.9.2 Illustration of Hyper-lattice Schema with Example

Here we draw the Hyper-lattice schema corresponding to Fig. 1.3. It consists of two lattices <A,B,C> and <B,C,D> which suggest the presence of two fact tables. The Hyper-lattice schema is shown in the Fig. 1.10.

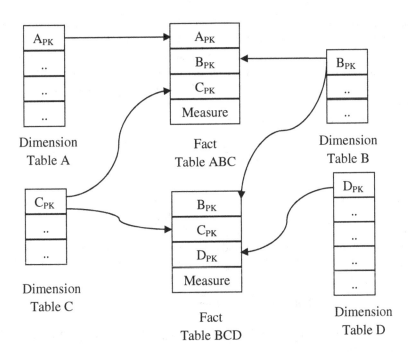

Fig. 1.10 An example of hyper-lattice schema

1.10 Using Hyper-lattice Towards Efficient Information Retrieval

Hyper-lattice consists of numbers of cuboids. Traversing these cuboids within Hyper-lattice is often subject of interest. Whenever roll-up operations are applied to move up in the Hyper-lattice or drill-down operation to move down in the structure more than one path could exist. In these types of operations, thus finding out an improved alternative in terms of both space and time lead to efficient resource utilization.

1.10.1 Algorithm to Find Efficient Space and Time Saving Path in Hyper-lattice Structure

In this work, cost factor is considered in terms of the size of every cuboid. The size of every cuboid is computed by multiplying the numbers of tuples in the associated cuboid with the size of the tuple. The size of every tuple is same. Size of the tuple is calculated by adding the length of each attribute pertaining to the cuboid and the size of the measure. The number of tuples in every cuboid is counted which is obtained from the inherent functions of the underlying relational database package.

Say a cuboid consists of the dimensions D_1, D_2,...,D_N. These dimensions are represented in the cuboids in the form of the attributes A_1, A_2,...,A_N. The size of the attributes A_1, A_2,...,A_N are S_1, S_2,...,S_N respectively. One or multiple measure is associated with each cuboid. However the measure values are same for every cuboid. Say the size of the measure is M. Consider that a cuboid consists of C tuples

The size of each cuboid is calculated as

$$C \times (\sum_{i=1}^{N} S_i + M)$$

The motivation of the work in this section is to find out an algorithm to identify a path between a pair of cuboids in Hyper-lattice which saves both space and time in terms of minimum amounts of size requirements.

The algorithm **SelectMinPath** finds the reduced path between two cuboids Source and Target by calculating the generation of minimum numbers of distinct values among the alternative cuboids.

Algorithm SelectMinPath (Source, Target)

Source is the starting cuboid where from the searching begins and Target is the destination cuboid.

Step 1:   (Array A and B stores the tagged values of the dimensions associated
          with Source cuboid and Target Cuboid respectively)
          A=SplitTagged(Source);
          B = SplitTagged(Target);
Step 2:   (Check whether path exists between Source and Target)
          If B is a subset of A
                  Tagged Value of Source is stored in Queue Q;
                  Set CUB=Source;

          else
                  Exit.
Step 3:   Repeat steps 3 to 10 until Target is reached.
Step 4:   (Initialize array C to store cuboids of upper level)
          Initialize the array C to NULL.
Step 5:   (Find out all the cuboids that could be generated in the immediate upper
           level from the current cuboid CUB)
          Generate the tagged Value of all the cuboids in the immediate upper level
          from the currently selected cuboid CUB and store in array C;
          /* Assume that the numbers of cuboids generated from CUB in the imme-
          diate upper level is NC */
Step 6:   (Initialize array D to store valid subset of cuboids from C which contain C)
          Initialize the array D to NULL.
Step 7:   (Find out those cuboids of C where from Target could be reached)
          For i = 1 to NC
                  T=SplitTagged(C[i]);
          If T contains tagged value of Target then
                  store cuboid C[i] in D.

Step 8:   (Choose the cuboid with minimum numbers of distinct values)
          Find out which cuboid CUB in D contains minimum numbers of distinct
          values.
Step 9:   Tagged Value of CUB is stored in Queue Q.
Step 10: (Check whether Target is reached)
          If CUB=Target
                  Break from loop;
          else
                  Goto Step 3.

Procedure SplitTagged(V)

Step 1:   Express the tagged value V of a cuboid as power of sum with base 2 on
          the tag numbers of the associated dimensions.
          $V = \Sigma\, 2^i$ for all i ϵ X:X is a subset of {0, 1, 2,..., N} for n dimensions
Step 2:   Store the tagged value of associated dimensions in an array T
Step 3:   Return T

Procedure SplitTagged(V)

Step 1:   Express the tagged value V of a cuboid as power of sum with base 2 on
          the tag numbers of the associated dimensions.

          $V = \Sigma\, 2^i$ for all $i \in X$:X is a subset of $\{0, 1, 2,..., N\}$ for N dimensions

Step 2:   Store the tagged value of associated dimensions in an array T

Step 3:   Return T

1.11 Lemmas on the Hyper-lattice

Lemma 1 *If a new dimension (with new tag number) is added to the lattice/
Hyper-lattice at level* $(i + 1)$, *a new base cuboid is formed at level i to accom-
modate the new dimension. The number of new dimension patterns that are added
to level* $(i + 1)$ *is* $^{n-i}C_{(n-i-1)} - 1$, *where n is the maximum level of the lattice /
Hyper-lattice.*

Proof In every upper level of lattice/Hyper-lattice, dimension of cuboid pattern
is reduced by 1 starting from the 0th level. If a new dimension is added at level
$(i + 1)$ the corresponding base cuboid is formed at level i. At level i number of
dimension of each cuboid is $(n - i)$, as at every upper level of cuboid dimension
is reduced by 1, hence at ith level dimension is reduced by i. At level $(i + 1)$ the
number of dimension is $(n - i - 1)$. So $^{n-i}C_{(n-i-1)}$ dimension pattern are generated
at $(i + 1)$-th level from the newly generated base cuboid.

   However, the dimension pattern to which the new dimension is actually added
that is already present at $(i + 1)$-th level. Thus the number of dimension patterns
that are actually being added at $(i + 1)$-th level is $^{n-i}C_{(n-i-1)} - 1$.

Lemma 2 *If a dimension is added at some dimension pattern at some level within
the Hyper-lattice which is already present (existing tag number) in the existing
Hyper-lattice, then the number of new dimension pattern that is being added at
that level is given by* (|Newly Generated Cuboid Pattern| − (|Newly Generated
Cuboid Pattern |∩| Previously Existing Cuboid Pattern|)). *The operator* | | *returns
the cardinality of a set.*

Proof Whenever a new dimension is added to a particular level of certain dimension pattern then a new base cuboid is created at level below with the new dimension and the existing dimension pattern of the point of insertion. It is required to
generate the all possible dimension patterns at immediate upper level from this
newly created base cuboid. It may be possible some of the dimension patterns
are already exist at immediate upper level. So there is no further need to generate
those existing patterns. The number of patterns that already exists can be given by
the following expression:

   (|*Newly Generated Cuboid Pattern*|∩|*Previously Existing Cuboid Pattern*|).

Thus the number of newly generated cuboid pattern is given by

($|Newly\ Generated\ Cuboid\ Pattern|$

$-$ ($|Newly\ Generated\ Cuboid\ Pattern|\ \cap\ |Previously\ Existing\ Cuboid\ Pattern|$)).

Lemma 3 *If a new dimension (with new tag number) is added to the lattice/Hyper-lattice at level i, the number of new dimension patterns that are being added is given by* ($2^{n-i+1} - 2^{n-i}$), *where n is the maximum level of the lattice/Hyper-lattice.* (*Assumption: new dimension could be added from 1 to n $-$ 1 level of lattice/Hyper-lattice*).

Proof If a dimension is added at ith level, it forms a new base cuboid at level $(i-1)$. At $(i-1)$th level, the number of dimension of every cuboid is $n-(i-1)$. With the newly formed base cuboid of $(n-(i-1))$ dimension it is possible to generate total 2^{n-i+1} dimension patterns in the Hyper-lattice. Now prior to the inclusion of this new dimension some of the dimension patterns were already present, which can be formed also from the newly formed base cuboid, but those are no longer required to generate. From $(n-i+1)$ dimensions, if the new dimension is excluded only $(n-i)$ dimensions are left. With the previously existing $(n-i)$ dimensions, it is possible to form 2^{n-i} dimension patterns, which already exists. Thus new dimension patterns are that are required to generate is given by $(2^{n-i+1} - 2^{n-i})$.

Lemma 4 If a dimension is added at some dimension pattern at some level i, within the Hyper-lattice of level n, which is already present (existing tag number) in the existing Hyper-lattice, then the number of new dimension patterns that are being added is given by ($\Sigma(i-1$ to $n-1)|$ Newly Generated Cuboid Pattern $|-(\Sigma(i-1$ to $n-1)\ |$ Newly Generated Cuboid Pattern $|\cap\Sigma(i-1$ to $n-1)|$ Previously Existing Cuboid Pattern $|$)) The operator $|\ |$ returns the cardinality of a set.

Proof Whenever a dimension (already present in Hyper-lattice structure) is added to a particular level i, and at a particular dimension pattern then a base cuboid is created at level $i-1$ with the new dimension and the dimension pattern at the point of insertion. At this point all possible dimension patterns are to be generated from the immediate upper level to $(n-1)$ level of the Hyper-lattice. Now some of the dimension patterns are already present in the Hyper-lattice structure. Thus from level $(i-1)$ to level $(n-1)$ all the dimension patterns are to be generated that are not existing in the Hyper-lattice structure. Thus the same formula of Lemma-2 is to be used and iterated from level $(i-1)$ to level $(n-1)$. So the result is ($\Sigma(i-1$ to $n-1)$ |Newly Generated Cuboid Pattern|$-(\Sigma(i-1$ to $n-1)$ |Newly Generated Cuboid Pattern|$\cap\Sigma(i-1$ to $n-1)$ |Previously Existing Cuboid Pattern|))

Lemma 5 *Whenever a new dimension is added into the lattice/Hyper-lattice structure the number of possible ways to visit to a particular cuboid from the immediate lower level is given by* (Current level $+$ number of new dimension added in the original lattice structure).

Proof A particular cuboid in a certain level, say k, of lattice can be reached from the immediate lower level in k ways. Now whenever a dimension is added in the Hyper-lattice at level k a base cuboid is formed at level $(k - 1)$. So that particular cuboid could be visited from the lower level in one more way due to the addition of new cuboid. So whenever a new dimension is added in the hyper lattice structure that cuboid could be visited. So addition of new n dimensions in the original lattice structure may result in addition of n new paths to visit a cuboid from the lower level. So the maximum numbers of ways visiting a cuboid from the immediate lower level is given by (Current level + number of new dimension added in the original lattice structure).

1.12 Example of Computational Benefit in Terms of Space and Time

Here, we are discussing about different aspects of computational benefits of Hyper-lattice. This new model is beneficiary in terms of less time for searching, lesser numbers of cuboids generation, capable to answer more queries than lattice.

1.12.1 Reduction of Computational Time During Searching

Hyper-lattice of Fig. 1.7 is considered and the numbers of distinct values for the dimensions A, B, C and D are 100, 200, 150 and 250 respectively. The problem is to find the path between the cuboids <A,B,C,D> and <C,D> which saves both space and time using the algorithm SelectMinPath.

Solution The cuboids that are generated from <A,B,C,D> to the immediate upper level are <A,B,C>, <A,B,D>, <A,C,D> and <B,C,D>. Among these four cuboids only two cuboids namely <A,C,D> and <B,C,D> contain the pattern <C,D>. Cuboid <A,C,D> contains $100 \times 150 \times 250 = 37,50,000$ distinct values, whereas cuboid <B,C,D> contains $200 \times 150 \times 250 = 75,00,000$ distinct values. Thus the algorithm SelectMinPath chooses the cuboid <A,C,D> in this step. The cuboids that are generated in the next level from <A,C,D>, are <A,C>, <A,D> and<C,D>. Thus according to the algorithm SelectMinPath selected path is <A,B,C,D> $\rightarrow$ <A,C,D> $\rightarrow$ <C,D>. If this algorithm is not applied the path <A,B,C,D> $\rightarrow$ <B,C,D> $\rightarrow$ <C,D> could be chosen, which requires the generation of higher numbers of distinct values over the selected path. Hence the reduction of space is achieved. Moreover the higher level is reached from lower level in Hyper-lattice using Roll-up operation which is equivalent to Group-by operation. It is known that the time requirement of Group-by is directly proportional to the numbers of distinct values in the data set. Hence this algorithm reduces computational time also.

1.12.2 Less Cuboids Generation Than Lattice

A lattice with n dimensions has 2^n cuboids. If a new dimension is added that mean forming a new lattice with $2^{(n+1)}$ cuboids. This means generation of cuboids $2^{(n+1)} - 2^n = 2^n$ cuboids. Where as if a new dimension is added in the lattice structure of n dimensions, it could be added only between the level 1 to $(n - 1)$. This would form a new base cuboid at 1 level below the target cuboid. Say, the level of new base cuboid is L. In this level the number of dimension of each cuboid is $(n - L)$. From this new base cuboid if a lattice is formed it would consists of $2^{(n-L)}$ cuboids. This value is clearly less than 2^n. Moreover among these $2^{(n-L)}$ cuboids only few (those cuboids which has the new dimension as member) are generated, remaining are already in the existing structure. Thus the use of Hyper-lattice leads to the generation of only fewer numbers of cuboids.

1.12.3 Higher Numbers of Query Answering

Say, initially a lattice consists of n dimensions. In case of the traditional lattice of cuboids even if newer information is available in terms of a new dimension starting from any intermediate cuboid, the same cannot be inserted into the lattice. However, this may be accommodated in a Hyper-lattice structure as proposed. Over a period of time, say k newer dimensions are inserted to form a Hyper-lattice starting from the lattice of n dimensions. It is obvious that Hyper-lattice could be used to trace the best possible path for execution of queries that involve one or more of these newer k dimensions, where as traditional lattice may be used only to handle the queries formed over the original n dimensions.

This theoretical observation is verified using a simulation experiment by executing random queries in a system based on lattice and Hyper-lattice structures. Four separate lattices are taken initially with 4, 6, 8 and 10 dimensions respectively. In each of these, an extra dimension is added to some cuboid other than the base cuboid of the respective lattice to create Hyper-lattices with 5, 7, 9 and 11 dimensions respectively. In the next step, random query sets are generated on both lattice and corresponding Hyper-lattices involving 1, 2 and 3 attributes in each query. In each query set, 20 queries are executed. As expected, Hyper-lattice could trace all the queries. However, the lattice fails for queries involving an attribute from the extra dimension added. In the graph (Fig. 1.11) the quantitative average performance of Hyper-lattice and lattice for the four sets are reflected.

X axis in the above figure indicates the dimensions for the Hyper-lattice structures and Y axis represents the success rate or the hit ratio of query for both the Hyper-lattice and lattice structures on a normalized scale of $0 - 1$. The third axis represents the different number of query attributes on which the test query sets are formed. In case of the all four series in X axis, the hit ratio for the corresponding Hyper-lattice is 1 where as the value of lattice varies and is less than 1. It may be

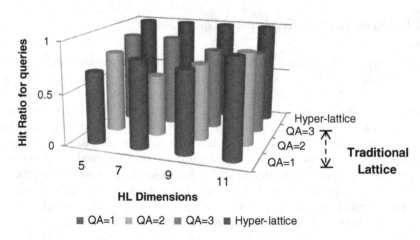

Fig. 1.11 Performance comparison between traditionnel lattice vis-à-vis hyper-lattice

noted here that even for the lattice it may be 1 if the query set does not involve any attribute from the newer dimension added over the traditional lattice structure. However, in our experiment, none of the randomly selected query sets are formed like that. The average performance of the proposed Hyper-lattice vis-à-vis the traditional lattice is shown in Fig. 1.11. In each case only addition of 1 dimension is considered. If the numbers of new dimensions are increased, the role of tracing queries using Hyper-lattice would further outperform the traditional lattice structure.

References

1. Sen S, Chaki N, Cortesi A (2009) Optimal Space and time complexity analysis on the lattice of cuboids using galois connections for data warehousing. In Proceedings of the IEEE 4th international conference on computer sciences and convergence information technology (ICCIT 2009), pp 1271–1275
2. Agarwal S, Agrawal R, Deshpande PM, Gupta A, Naughton JF, Ramakrishnan R, Sarawagi S (1996) On the computation of multidimensional aggregates. In: Proceedings of the 22th international conference on very large data bases (VLDB), pp 506–521
3. Halder R, Cortesi A (2012) Abstract interpretation of database query languages. J Comput Lang Syst Struct 38(2):123–157
4. Chen Y, Dong G, Han J, Wah BW, Wang J (2002) Multi-dimensional regression analysis of time-series data streams. In: Proceedings of the 28th international conference on very large data bases (VLDB), pp 323–334
5. Gray J, Chaudhuri S, Bosworth A, Layman A, Reichart D, Venkatrao M, Pellow F, Pirahesh H (1997) Data cube: a relational aggregation operator generalizing group-by, cross-tab, and sub-totals. Data Min Knowl Discov 1(1):29–53
6. Han J, Kamber M (2006) Data mining: concepts and techniques, 2nd edn. The Morgan Kaufmann series in data management systems, Elsevier Science. ISBN 13: 978-1-55860-901-3
7. Li X, Han J, Gonzalez H (2004) High-dimensional OLAP: a minimal cubing approach. In: Proceedings of the 30th international conference on very large data bases (VLDB), pp 528–539

8. Papadias D, Kalnis P, Zhang J, Tao Y (2001) Efficient OLAP operations in spatial data warehouses. In: Proceedings of the 7th international conference on advances in spatial and temporal databases (SSTD). Springer, LNCS 2121, pp 443–459

9. Wang M, Iyer B (1997) Efficient roll-up and drill-down analysis in relational databases. In: Proceedings of the 1st SIGMOD workshop on research issues on data mining and knowledge discovery, pp 39–43

10. Wijsen J, Ng RT, Price D (1999) Discovering roll-up dependencies. In: Proceedings of the 5th ACM SIGKDD international conference knowledge discovery and data mining, pp 213–22

11. Harinarayan V, Rajaraman A, Ullman JD (1996) Implementing data cubes efficiently. In: Proceedings of the 22nd ACM SIGMOD international conference on management of data, pp 205–216

12. Baralis E, Paraboschi S, Tenient E (1997) Materialized view selection in a multidimensional database. In: Proceedings of the 23rd international conference on very large database (VLDB), pp 156—165

13. Ullman JD (1996) Efficient implementation of data cubes via materialized views. In: Proceedings of the 2nd international conference on knowledge discovery and data mining (KDD), pp 386–388

14. Yu JX, Lu H, (2001) Multi-cube computation. In: Proceedings of the 7th international conference on database systems for advanced applications, pp 126–133

15. Dehne F, Eavis T, Hambrusch S, Rau-Chaplin A (2001) Parallelizing the data cube. In: Proceedings of the 8th international conference on database theory (ICDT), pp 129–143

16. Dehne F, Eavis T, Hambrusch S, Rau-Chaplin A (2006) The cgmCUBE project: optimizing parallel data cube generation for ROLAP. J Distrib Parallel Databases 23(2):99–126

17. Han J, Pei J, Dong G, Wang K (2001) Efficient computation of iceberg cubes with complex measures. In: Proceedings of the 27th ACM SIGMOD international conference on management of data, pp 1–12

18. Zhao Y, Deshpande P, Naughton JF (1997) An array-based algorithm for simultaneous multidimensional aggregates. In: Proceedings of the 23rd ACM SIGMOD International conference on Management of data, pp 159–170

19. Chen Y, Dehne F, Eavis T, Rau-Chaplin (2004) A building large ROLAP data cubes in parallel. In: Proceedings of the 8th international database engineering and applications symposium (IDEAS), pp 367–377

20. Beyer K, Ramakrishnan R (1999) Bottom-up computation of sparse and iceberg cubes. In: Proceedings of the 25th ACM SIGMOD International conference on management of data, pp 359–370

21. Hu XG, Wang DX, Liu XP, Guo J, Wang H (2004) The analysis on model of association rules mining based on concept lattice and a-priori algorithm. In: Proceedings of the 3rd international conference on machine learning and cybernetics (ICMLC), pp 1620–1624

22. Xin D, Han J, Li X, Wah BW (2003) Star-cubing: computing iceberg cubes by top-down and bottom-up integration. In: Proceedings of the 23rd international conference on very large database (VLDB), pp 476–487

23. Shao Z, Han J, Xin D (2004) MM-cubing: computing Iceberg cubes by factorizing the lattice space. In: Proceedings of the 16th international conference on scientific and statistical database management (SSDBM), pp 213–222

24. Xin D, Han J, Li X, Shao Z, Wah BW (2007) Computing iceberg cubes by top-down and bottom-up integration: the starcubing approach. IEEE Trans Knowl Data Eng 19(1):111–126

25. Barbara D, Wu X (2000) Using loglinear models to compress datacube. In: Proceedings of the 1st international conference on web-age information management (WAIM), pp 311–322

26. Vitter JS, Wang M, Iyer BR (1998) Data cube approximation and histograms via wavelets. In: Proceedings of the 7th international conference on information and knowledge management (CIKM), pp 96–104

27. Jinghua H, Mei Y, Xiaowei L, Xinna S (2010) The design and implementation of MDSS based on data warehouse. In: Proceedings of the 1st international conference on computing, control and industrial engineering (CCIE), pp 42–45
28. Huang C, Zeng Z, Yue D (2006) The design and its achieving method on multi-dimension data warehouse of medical waste management. In: Proceedings of the 1st IEEE international conference on service operations and logistics, and informatics, pp 873–877
29. Tripathy A, Mishra L, Patra PK (2010) A multi dimensional design framework for querying spatial data using concept lattice. In: Proceedings of the 2nd international advance computing conference (IACC), pp 394–399

Chapter 2
Applications of Hyper-lattice in Real Life

In this chapter, we illustrate different business and real life situations where Hyper-lattice increases efficiency in data analytics. In Sect. 2.2, different case studies are considered on five different application environments. Within a short space, we have considered variations common in data warehouses and how to use Hyper Lattice to represent, and model such situations.

2.1 Introduction

Data warehouse is widely used across the organizations for different business applications. Every organization today gives higher priority on business intelligence for better business planning. We discuss applications of Hyper-lattice considering a few case studies for diverse business environment.

2.2 QoS Parameters in Service Oriented Systems

In this use case, we show the value of the proposed Hyper Lattice model where data for analytics is partially available on a particular dimension.

In a service oriented system QoS (Quality of Service), parameters are equally important along with the functional requirement of the services. These parameters could be treated as dimensions for data warehouse applications. The commonly used dimensions are Availability, Portability, Modifiability, Reliability. Hence based on these 4 dimensions a lattice of cuboids could be formed, which consists of $2^4 = 16$ cuboids.

Say, over the time it has become important to incorporate Security in this application. However, it's seen that detailed data required to introduce Security is

© The Author(s) 2016
S. Sen et al., *Hyper-lattice Algebraic Model for Data Warehousing*,
SpringerBriefs in Applied Sciences and Technology,
DOI 10.1007/978-3-319-28044-8_2

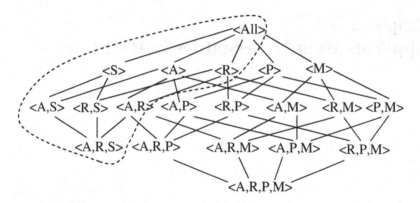

Fig. 2.1 Hyper-lattice of QoS parameters

not available at level 0, while data for Security is there against the variations in Availability and Reliability.

In order to incorporate Security in analytics in this context, a new base cuboid may be formed as <Availability, Reliability, Security> in level-1 of the proposed Hyper-lattice. Thereafter in the next level, 2 cuboids would be formed as <Availability, Security>, and <Reliability, Security>. After this <Security> cuboid would be formed in level-3. Finally, the aggregation will be done in apex level at <All>.

In Fig. 2.1, the newly added lattice is shown (marked within dotted curve) which makes the structure Hyper-lattice. In the figure Availability, Reliability, Portability, Modifiability Security are noted as A, R, P, M and S.

Advantage Derived from Hyper-lattice:

In this use case we assume that the variations in Security data for different values of Modifiability and Portability are not in hand. Consequently, in case of a classical Lattice structure, if we try to enforce a new dimension for Security at level 0, there would be NULL entries in a large number of cuboids.

In this example, the Hyper-lattice is required as Security data is available with only two of the dimensions among the four initial dimensions. This example illustrates the use of Hyper-lattice in Service Engineering.

Here, the initial lattice with 4 dimensions had $2^4 = 16$ cuboids. We have added a 5th dimension to it. In case of a classical lattice, this implies existence of $2^5 = 32$ cuboids in the structure. Also the maximum path length from the base cuboid to the apex will be increased. In the Hyper Lattice in Fig. 2.1, only 4 new cuboids are added to effectively consider variations for the newly introduced dimension Security at all appropriate levels.

This means we do not need to store 12 additional cuboids those contain null values only. Hence we save storage of 37.5 % of cuboids. It may also be noted here that in this case, 4 of the cuboids from the initial lattice are overlapping the new lattice (within the Hyper Lattice created) with <A,R,S> as the base cuboid.

It is important to consider here that the growth in the lattice structure with insertions of new dimensions at the base level is exponential in nature. However, adding a new dimension in a lattice to form a Hyper Lattice involves much less overhead. We discuss this in the rest of this case study.

Let's consider that a new dimension D is added K-levels below the Apex cuboid in an existing lattice of N dimensions, and therefore with $C1 = 2^N$ cuboids in it. This results in formation of a Hyper Lattice with two base cuboids. The lattice with this new base cuboid will have C2 cuboids as:

$$C2 = 2^{P-1}, \quad \text{where P = Maximum level } - \text{ K}$$

However, there will be C3 elements common to both the lattices in the newly created Hyper Lattice, as:

$$C3 = 2^{P-2}, \quad \text{where P = Maximum level } - \text{ K}$$

Hence, the newly created Hyper Lattice will have only $C2 - C3 = 2^{P-2}$ new elements in it. This is much less than 2^{N+1}, the number of cuboids required in the new lattice for (N + 1) dimensions. This reduction in algebraic structure will be different depending on two factors:

- Dimension of the lattice
- Level at which insertion of the new base cuboid is made

We have analyzed and verified that there would be a variation from 37.25 % to nearly up to 50 % in terms of number of cuboids less in a Hyper Lattice with two base cuboids as compared to the immediate higher dimension of lattice. Table 2.1 presents some data on this. We have ignored insertion at just one level below the apex level as well as in level 0, as these are trivial cases.

Figure 2.2 presents a comparative analysis in the growth for number of cuboids in the two algebraic structures. We have taken data from Table 2.1 for insertion at

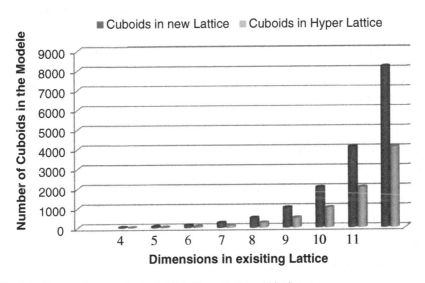

Fig. 2.2 Comparative growth of cuboids in Hyper-lattice and lattice

Table 2.1 Cuboids in lattice versus Hyper lattice

Existing dimensions (N)	Cuboids in new lattice	K	P	Number of cuboids				Savings in percentage (%)
				C2	C3	New in HL	Total in HL	
4	32	1	4	8	4	4	20	37.50
4	32	2	3	4	2	2	18	43.75
5	64	1	5	16	8	8	40	37.50
5	64	2	4	8	4	4	36	43.75
5	64	3	3	4	2	2	34	46.88
6	128	1	6	32	16	16	80	37.50
6	128	2	5	16	8	8	72	43.75
6	128	3	4	8	4	4	68	46.88
6	128	4	3	4	2	2	66	48.44
7	256	1	7	64	32	32	160	37.50
7	256	2	6	32	16	16	144	43.75
7	256	3	5	16	8	8	136	46.88
7	256	4	4	8	4	4	132	48.44
7	256	5	3	4	2	2	130	49.22
8	512	1	8	128	64	64	320	37.50
8	512	2	7	64	32	32	288	43.75
8	512	3	6	32	16	16	272	46.88
8	512	4	5	16	8	8	264	48.44
8	512	5	4	8	4	4	260	49.22
8	512	6	3	4	2	2	258	49.61
9	1024	7	3	4	2	2	514	49.80
10	2048	8	3	4	2	2	1026	49.90
11	4096	9	3	4	2	2	2050	49.95
12	8192	10	3	4	2	2	4098	49.98

level 1 always ($K = N - 2$) so as to illustrate the maximum differences between the two algebraic structures.

We avoid repeating similar discussion at the end of each Use Case as the analysis and the results above are generic.

2.3 Sales Hyper-lattice

In this use case, we show the value of the proposed Hyper Lattice model where even if data for analytics is completely available for all the dimensions, analysis is required only on a sub-set for a particular dimension.

Case 2.3A

Let's assume that in a business application environment there are four dimensions say Time, Location, Product and Supplier. With these four dimensions there would be 2^4, that is, 16 cuboids in the lattice of cuboids. Later the organization finds out

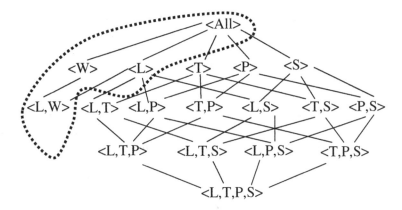

Fig. 2.3 Sales Hyper-lattice

that business varies due to the sudden change of the weather in its different business locations. Thus, it needs to incorporate the change of weather information in the data warehouse. However, the organization does not have the data to identify the relationship between Weather with wither of the Product, Time or Supplier dimensions.

In the classical model of lattice it is not possible to incorporate Weather as a dimension. Hence the organization would fail to make proper business decisions as the analysis lacks weather related information.

Hyper Lattice allows it to form a new cuboid consisting of Location and Weather. The dimension weather can be represented based on the rainfall, temperature variation etc. In the Hyper-lattice, we can add a new base cuboid <Location, Weather> two levels below the apex element and from it the cuboid <Weather> is generated in one level lower than the apex cuboid (Fig. 2.3).

Here, even if the organization has the information of Weather along with all the four dimensions, the new business analysis is required only on Weather and Location. In the lattice model of cuboids, if the Weather dimension is included, then the number of cuboids would be simply double and makes it a $2^5 = 32$ cuboids structure. However, the analysis is required only on variation of Weather for different business Locations.

Advantage Derived from Hyper-lattice:

Using the Hyper-lattice model we would add only 2 cuboids <Location, Weather> and <Weather> in the existing structure of $2^4 = 16$ cuboids.

This means that we need not store 14 additional cuboids containing only NULL values for the Weather dimension. Hence, we save storage of 43.75 % of cuboids. It may be noted here that in this case, 2 of the cuboids from the initial lattice are overlapping the new lattice (within the Hyper Lattice created) with <Location, Weather> as the based cuboid.

Case 2.3B

Now the previous case is extended further. Say, the organization got information on Product dimension along with the Weather and Location dimension. Then it is possible to perform the analysis on these 3 dimensions all together. This is going to form a new base cuboid <Product, Location, Weather>. Now from this new base cuboid in Hyper-lattice only the new cuboid that is to be generated is <Product, Weather>. It is to be noted <Location, Weather> does not remain a base cuboid.

Advantage Derived from Hyper-lattice:

If we combine Case 1A and Case 1B the numbers of cuboids in the Hyper-lattice would be 19 and could avoid 32 cuboids of lattice structure. This means we do not need to store 13 additional cuboids those contain null values only. Hence we save storage of 40 % of cuboids.

2.4 Share Market Hyper-lattice

In this use case, we show the importance of the proposed Hyper Lattice model where a large numbers of dimensions are participating but new dimensions are inserted at some of the cuboids for very specific analysis.

Case 2.4A

We consider another interesting case here. Here we discuss about share market. There are lot of issues which drive share market. Here we initially assume some of the dimensions like Time (T), Interest rate of bank (R), Average profit-loss of the companies enlisted in the share market (L), Price of petrol/diesel (P), Gold Price (G), Government Decision (D). Thus to start with we have 6 dimensions and therefore there would be 64 cuboids to start with, in the lattice of cuboids. Government Decision (D) dimension may be categorized in different perspective. In a country like India percentage of privatization of nationalized companies or the percentage of FDI (Foreign Direct Investment) for certain business sectors plays important role in share market. These decisions are taken by the Government. In this way Government Decision (D) dimension may be formed. The base cuboid is <T,R,L,P,G,D>. Now over the time it is found that the Industry/Corporate decisions (I) is also playing important role in share market. Industry/Corporate decisions (I) can also be of different types such as acquiring new companies or investment in foreign locations etc. This new dimension is related with the cuboid <R,L,P,G> and thus the information is present in terms of <R,L,P,G,I>. Hence this dimension I would be added to <R,L,P,G> to form a new base cuboid <R,L,P,G,I> in level-1 of the Hyper-lattice. New cuboids would be created until 1-dimensional cuboids are created.

After the creation of base cuboid <R,L,P,G,I> in level-1 new 4D cuboids would be generated in level-2. These new cuboids in level-2 are <R,L,P,I>, <R,L,G,I>, <R,P,G,I> and <L,P,G,I>. In level-3, 6 new cuboids would be generated namely <R,L,I>, <R,P,I>, <L,P,I>, <R,G,I>, <L,G,I> and <P,G,I>. After this in next level (level-4) 4 new cuboids would be created namely <R,I>, <L,I>, <P,I> and <G,I>. Finally in level-5 one new cuboid would be formed as <I>.

Advantage Derived from Hyper-lattice:

In this Hyper-lattice, we identify the complex business scenarios of share market. In order to start with numbers of dimensions are already high. Later new dimensions are added at specific level. Normal lattice does not allow the insertion of dimension at any other level except the base level and this increase the numbers of cuboids exponentially. Hyper-lattice ensures that the cuboids are not increased exponentially.

We find here that in order to meet the requirement, 15 cuboids are added in the Hyper-lattice. If we don't have the Hyper-lattice we need to create a separate lattice of cuboids starting from the base cuboid <R,L,P,G,I>. This new lattice of cuboids would have number of cuboids overlapping with the existing lattice of cuboids and therefore cause redundancy of data and wastage of space.

On the other hand if we consider that we have all the relevant information about the dimension I along with <T,R,L,P,G,D> then this would create a new lattice of cuboids by adding 64 new cuboids, whereas the proposed Hyper-lattice model serves the requirement by adding 15 cuboids only.

Case 2.4B

Here we consider another situation. It is found that the political condition of a country is also affecting the share market. Hence we need to consider Political Condition (C) as one of the dimension. Political Condition (C) may be defined in terms of change of government, stability of government etc. This dimension C, is affecting the dimensions Price of petrol/diesel (P) and Gold Price (G). Hence a new base cuboid is formed in the form of <P,G,C> at level-3 of Hyper-lattice. We need to generate the cuboids <P,C> and <G,C> at level-2 and then <C> at level-1.

Advantage Derived from Hyper-lattice:

This example again shows instead of doubling the structure from 6D lattice of cuboids to 7D lattice of cuboids only 4 cuboids are added to satisfy the requirement. This means instead of adding 64 cuboids in this context, only 4 cuboids to be added.

It may be noted here that the analysis on share market may be very complex. Consider in some country the share market is driven by 20 issues. So in this case the lattice of cuboids would consist of 2^{20} cuboids. Say 4 new dimensions are introduced separately at level-19 with 4 different cuboids. This is going to add only 4 cuboids in the Hyper-lattice. Whereas if we have to use lattice then the numbers of cuboids would have been $2^{24} = 1,67,77,216$. Using Hyper-lattice we have only $10,48,584$ ($2^{20} + 8$) cuboids. This means we could save $1,57,28,632$ cuboids. In terms of asymptotic notation instead of $O(2^{24})$ we are getting $O(2^{20})$ only.

2.5 Health Monitoring System Hyper-lattice

In this use case, we show the importance of the proposed Hyper Lattice model where some of the closely related dimensions are participating but analysis is required for specific cuboids.

A healthcare system currently consists of four dimensions namely Time, Calories Burned, Heart Rate and Blood Pressure. As this is a four dimensional structure it consists of 16 cuboids to form the lattice. Over the time Sugar Level of the patient is available. But it is available along with Heart Rate and Blood Pressure as whenever sugar testing is done Heart rate and Blood pressure are being checked. Note that the abstraction on Time dimension that is present in database is different than of sugar testing interval. Hence, if in the current warehouse Time dimension is in the form of day and sugar testing is done weekly then we can't include Time dimension with Sugar level for the analysis. Moreover as Calories burned is not checked during sugar testing this dimension could not appear in the analysis. Hence we could form the base cuboid with the dimensions Sugar Level, Heart Rate and Blood Pressure.

In this example, the benefit of using Hyper-lattice is that instead of maintaining two lattice of cuboids

(i)   <Time, Calories Burned, Heart Rate, Blood Pressure>
(ii) <Sugar Level, Heart Rate, Blood Pressure>,

we could work with one Hyper-lattice. We need to generate the cuboids <Sugar Level, Heart Rate>, <Sugar Level, Blood Pressure> and <Sugar Level> in the Hyper-lattice.

Advantage Derived from Hyper-lattice:

Health related information is very crucial and critical also in some case. Thus if same information is saved separately in different places there may be a chance of inconsistency of data. This is avoided here by sharing the overlapping cuboids such as <Heart Rate, Blood Pressure>, <Heart Rate>, <Blood Pressure>.

We took a small example in terms of numbers of dimensions. But in medical science new dimensions can be added any time based on the nature of disease. Obviously the disease and their symptoms (to be represented as dimension here) can increase rapidly. However some of the dimensions are irrelevant for some of the disease. Traditional lattice would create a huge numbers of cuboids those have no meanings but increase the structure exponentially.

2.6 Hyper-lattice for Traffic Management System

In this use case, we show the importance of the proposed Hyper Lattice model where one of the specific cuboid is very important and it is considered multiple times for insertion of new dimensions.

A traffic management system for a city is currently based on three dimensions namely Numbers of vehicles, People moving and hour which correspond to time. Based on three dimensions fact table is being built up to detect the speed of the vehicles. Hence the speed of the vehicle is to be considered as the measure. This system consists of 8 cuboids in the lattice of cuboids where the base cuboid is <Vehicle, People, Hour>.

In addition to this it is identified the climate condition also plays an important role. Here climate is important in two aspects. One is Rain and another is Fog. The quantity of Rain and Fog density is considered here. These two factors are separately related with <Vehicle, People>. Hence two new base cuboids would be formed <Vehicle, People, Rain> and <Vehicle, People, Fog>. The new cuboids that are to be generated are from these two base cuboids are <Vehicle, Rain>, <People, Rain>, <Rain> and <Vehicle, Fog>, <People, Fog>, <Fog>. Hence numbers of new cuboids generated are 8 including the two new base cuboids. Whereas if these two dimensions would have been added to the original base cuboid, then the new lattice of cuboids would have 32 cuboids. Hence instead of adding 24 cuboids we need to add only 8 cuboids to the Hyper-lattice.

Advantage Derived from Hyper-lattice:

This example shows higher level of overlapping of cuboids. The same cuboid has been chosen for insertion of new dimension twice. Hence the created Hyper-lattice in this example allows even further reduction in cuboid generation. This type of scenario where same cuboid is chosen multiple times for insertion, Hyper-lattice is more effective in terms of less cuboids generation.

2.7 Conclusions

In the five different use cases, we have shown diverse possible usage of Hyper Lattice as a new algebraic structure and its advantage over traditional lattices.

In the above case studies we have discussed about different dimensions. All the dimensions are represented in one abstracted form only. However, during analysis quite often data on these dimensions are required in different granularities. As for example in Sales Hyper-lattice, dimension Time may be represented in the form of Month, whereas for Traffic Management system the dimension Time is required in the form of Hour. Now it may be possible that in some week due to some festival, the sales is very high for a company compared to the other weeks. So for that week the company may need to analyze the business for the week instead of month. Therefore, if lattice of cuboids is integrated with the concept hierarchy of the dimensions we would analyze the business in a more customized way. This has been discussed in Chap. 3 to follow.

Chapter 3
Generating Co-operative Queries Over Concept Hierarchies

In this chapter, we discuss how using co-operative query languages can be used to increase efficiency of analytics when it's used on Concept Hierarchies. Concept Hierarchy presents the information of a same dimension in different abstracted levels. This abstraction allows us to identify the same data in multiple granularities and from different users' perspectives. Conventional query execution retrieves information in one abstracted form only for the given dimension. Actually traditional database management models including RDBMS do not store the concept hierarchy information. This would be more relevant for online analytic processing (OLAP) on data warehouse. Indeed, it is a challenge for designer of data analytics application software to use a query language to take the benefit of concept hierarchy towards extracting optimized information for specific users. In the rest of the chapter, we have explored cooperative query language in this context and establish its suitability.

3.1 Introduction

Representation of a dimension is an important decision in data warehouse. A dimension could be represented in different abstract forms. It is defined as a sequence from lower level detailed representations to higher level aggregated representations. This is known as concept hierarchy. Concept hierarchy could be in the form of total order (like in Fig. 3.1) or partial order (like in Fig. 3.2).

This type of situation defines the requirements of cooperative queries. A cooperative query is represented as a collection of number of queries that are logically cohesive. This helps in retrieving more information from database which is otherwise not possible using single query only.

© The Author(s) 2016
S. Sen et al., *Hyper-lattice Algebraic Model for Data Warehousing*,
SpringerBriefs in Applied Sciences and Technology,
DOI 10.1007/978-3-319-28044-8_3

Fig. 3.1 Concept hierarchy for total order

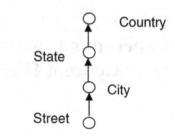

Fig. 3.2 Concept hierarchy for partial order

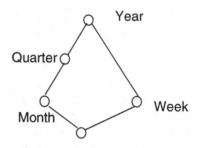

Cooperative querying [1] is a special type of Information Retrieval (IR) [2]. Cooperative query answering concepts are generally based on "human intelligent responses". Human tends to reply to the questions with informative or meaningful responses rather than reject answers, in the circumstances where the answers to the questions are either not known or negative. However, a typical database system on the other hand, does not have this essential cooperative answering capability. When a user submits a question in the form of a Boolean query, a standard database system usually replies with only "yes" or "no". This is because the conventional query answering system requires "exact matching" between the given query conditions and the corresponding answer properties. This exact match property of a typical query answering mechanism sometime causes user frustration, especially when a negative answer is not expected.

As mentioned in the concluding comments of Chap. 2, if lattice of cuboids is integrated with concept hierarchy of the dimensions, lot of flexibility is added to the domain of business analytics. However lattice of cuboids with concept hierarchy is a huge structure. It is given in the following formula

$$\prod_{I=1}^{N}(L_I + 1)$$

where, L_I is the numbers of levels associated with dimension I.

As for example, if in a system there are 3 dimensions namely Time, Location and Product; where the Time is abstracted as Day, Month and Year; Location is abstracted as City and State then the total numbers of cuboids in the lattice would be 24. This gives an idea how the numbers of cuboids grow rapidly in the lattice if the dimensions maintain concept hierarchies. In order to avoid these huge numbers of cuboids, generally the lattice of cuboids along with the concept hierarchy is not being stored.

Many [1, 5, 6] cooperative query answering techniques have been developed over the time to improve information retrieval by incorporating human-like intelligent question answering into the standard database systems. The common objective of cooperative querying systems is to improve system-user interactions. Cooperative querying gives database retrieval systems a human intelligence [1] by mimicking their ability to produce correct or informative answers. This is achieved by applying some artificial intelligence mechanisms, framing of rules or facts, from the existing and/or supplementary knowledge developed by application domain experts. Hence query generation [3] for different types of databases where the underlying data resided in other formats is potentially a challenging area of research.

Relational model is one of the most widely used data model [4] for database and SQL (Structured Query Language) is the corresponding query language and is used worldwide for different types of data centric applications. In this research work the focus is on generating Co-Operative query answering for relational model and the queries are formulated using SQL. Co-Operative query processing includes three major concepts like Query Refinement, Query Abstraction and Query Relaxation.

Query Refinement [5]

Query refinement is a process of converting a query Q to specific query representations Q1, Q2, ..., Qn according to the type abstraction hierarchy, through the following steps.

(a) Finding out the set of subtype objects through type specialization rewrites.
(b) Provide conversion mechanism between the attributes and the corresponding types.
(c) Based on each subtype, use type specialization to transform attributes referred to in the query to those in the subtype object.
(d) Based on each subtype, use both "type specialization rewrite" and "term specialization rewrite" to transform conditions referred to in the query to those related to the subtype object.

Query Abstraction [5]

Query abstraction is a process of converting a query Q into a more abstract representation, Q', through the following steps.

(a) Find the appropriate supertype object through type generalization rewrite.
(b) Convert the attributes to the corresponding types, and vice versa.
(c) Transform attributes referred to in the query to that of the supertype object through type generalization rewrite.
(d) Transform conditions referred to in the query to those related to the supertype object through both type generalization rewrite and term generalization rewrite.

Query Relaxation [6]

Traditional query processing models were based on the assumption that "user knows what he/she wants" and consequently he/she is able to precisely formulate the query. However, in real life users could query data without a deep knowledge of how the data are structured and they may not be aware of how to formulate meaningful queries from it. This situation has led current query processing models to change from the traditional belief into the new belief that "the user has an idea of what he/she wants" and the system has to automatically lead him/her to formulate meaningful queries by relaxing query constraints. This is referred to as query relaxation.

3.2 Problem Definition

Here we formally define the problem. In a data warehouse, one or more dimensions may have a concept hierarchy in different granularity. Thus, data on a particular dimension is represented in multiple abstractions in accordance with the underlying concept hierarchy. These abstractions are logically implied in the data warehouse. However, all the abstractions are not physically stored. In majority of the cases, only one form of the abstracted value is maintained for each dimension. Thus when a user query requires the information in multiple abstracted level, a single SQL query could not generate the answer. In this type of scenario separate queries are required to fetch the data partially.

However, these queries should be logically co-related so that collectively it generates the desired result. This has been incorporated here using co-operative query answering. As this is designed for SQL queries, all the underlying data are stored in tables. According to the proposed methodology, whenever any searching or traversing is required, corresponding SQL queries are generated. Queries that are successful in terms of generating the desired data are combined together to prepare the co-operative query answering.

Besides, the concept hierarchy is essentially a hierarchical structure. Hence, to fetch the data from this hierarchical structure, efficient traversal mechanism is required. This research work proposes a traversal mechanism to exploit the knowledge generated from roll-up (bottom-up traversal) and drill-down (top-down traversal) on concept hierarchy. There could be multiple solutions generated during traversal within the structure. Checking each of these towards finding the optimum solution has a huge overhead. In order to avoid this, a rank based system is incorporated. A rank [7] is pre-assigned to each data item. This rank could be calculated for each data item from the historical data associated with these data items by applying statistical methodology or analytical processing. This helps to filter the relevant data from a large set of available alternatives.

The objective of this research work is, first to find suitable traversal mechanism within the concept hierarchy with the knowledge of rank of associated data items. Second, we generate the equivalent co-operative query over the database or data warehouse to execute the traversal as defined in the database.

3.3 Proposed Methodology of Information Fetching from Concept Hierarchy Using Co-operative Query Language

In this methodology a normal query is partitioned into multiple queries to retrieve more information which is not possible by executing a single query only. This is performed here by using the concept of Co-operative query answering. This introduces several concepts like query relaxation, query filtering, query abstraction. On the other hand in a data warehouse an attribute could be represented in different abstract level in terms of concept hierarchy. It may be possible that the knowledge that the system want to infer, is not present in target level of concept hierarchy. Thus it is often required to traverse the structure of concept hierarchy to retrieve the desirable information from the appropriate level. Moreover to expedite the searching process and also to find out related information the concept of rank is associated with every item defined in concept hierarchy. The novelty of this research work is integrating the concept of Co-Operative query answering with concept hierarchy and also incorporating the concept of priority (rank) among the elements of concept hierarchy to search the elements efficiently and henceforth generating the corresponding Co-operative query answering.

3.3.1 Different Terminologies Proposed

In our proposed methodology we have used different data structure and terms which have specific meaning as well as specific purposes. They are explained here.

Level Number: Concept hierarchy is a hierarchical structure. Each item in this hierarchical structure is associated with a Level Number. The items lowest in the hierarchy, that is, where the data is at most granular level in Level Number 1. As moves up in the hierarchy Level number is increased by 1.

Concept Hierarchy Table (CHT): In this table—Level Number, Item Id, Item Name, Immediate Upper Level Item Id and Rank value of a particular item is assigned, which helps in traversing the concept hierarchy. CHT is defined for each level in the concept hierarchy—for each level only one CHT is assigned. Level Number is 1 for the most detailed data of concept hierarchy and highest for the most aggregated data of concept hierarchy. The structure of the table is given below:

Item Id	Item Name	Level Number	Immediate Upper level Item Id	Rank

According to application separate concept hierarchy tables could be maintained for separate levels. In this case the level number column could be eliminated from the corresponding table.

Search Table: This is an optional table. If the system uses multiple concept hierarchy tables this table could be used, otherwise ignored. It consists of Item Id, Item Name, and Level Number of a particular item. This optional table is used to reduce the search time by avoiding searching in multiple concept hierarchy tables. The structure of the table is given below:

Item Id	Item Name	Level Number

The entries of the Search Table are already present in the Concept Hierarchy Table, but it is used to save the searching time.

Connection Table: This table shows the connectivity of each item at level 1 with other items in that level. The structure of the table is given below:

Connecting Item Name	Connecting Item Id

Source: It is considered as the place from which the connection is to be found.

Destination: It is considered as the place up to which the connection is to be found.

Item Id: It is a unique id given to each item.

Rank of an item: Rank of the item is a value given to that item according to its priority in that particular table.

3.3.2 Assumptions

1. The system may or may not use Search Table. It depends upon the numbers of CHT. If more than one CHT is used, then it defined for each level in the concept hierarchy. Because searching in multiple CHTs is inconvenient. But if there is only one general table for the whole concept hierarchy then no need to create Search Table. In that case, the Search table is sorted by level number.
2. The levels of the Concept Hierarchy and assigned Ranks Vare predefined.
3. The connection table is defined for each item in Level Number 1 to show the connection with other items in Level Number 1. So for establishing connection the control has to start from Level Number 1. If necessary the control goes to the upper levels.

3.3.3 Algorithm Description

This algorithm traverses the given concept hierarchy in both the ways to generate the desired result. The traditional approach finds the solution only from the single node of the given concept hierarchy. Here the concept of traversing the structure helps in finding more information from the multiple nodes. Moreover the rank associated with each item of the abstraction of the concept hierarchy helps to reduce the search space. During the execution of algorithm whenever the searching is true corresponding query is generated. All of these queries generated are clubbed according to the sequence of generation to form the co-operative query language. Co-operative query language helps here to generate new information which is not possible otherwise using traditional SQL only.

The execution of the algorithm starts by checking whether the given input is in the Search Table (If the system uses it). The table contains all the items along with "Item Id" and "Level Number". If the user gives wrong input, an error message is displayed. This is done if multiple CHTs exist in the system. In the absence of multiple CHT only a generalized CHT is used for all the levels and no Search Table is required. In this context, the validation will be done within this generalized CHT only. Next the source and destination value will be retrieved from the search table or from the generalized CHT.

If it is found that both source and destination are in the Level Number-1 then the system tries to find direct connection between them otherwise the control will traverse to the Level Number 1 along the concept hierarchy by decreasing the level number. However one consideration is taken care. Every Item in the system is assigned a rank. The item which has the highest rank is selected from the lower level for source and destination in each case. In this way searching will continue until it reaches the target level.

After this, the connection between the source and destination is to be established. At first, the control searches for direct connection. If it gets a connection then the corresponding connection is given as output to the user. However, if it fails to get a direct connection then the system will try to find the connection through intermediaries. Here the rank of a particular item plays an important role in getting the indirect connection. At first the system selects the items with same rank as the Source (SRC) and Destination (DSTN) and tries to find out the connection among them from CHT and Connection Table. If connection found, it is displayed but in case of failure the items with higher rank values are considered. The searching is continued, if found appropriate message is given otherwise the items with lower rank are selected and each time the system tries to find the connection among the items. If at any stage connection is found then the algorithm would terminate by displaying the proper connection.

This may happen that from the above approach, the system could not find any connection then the system tries to find connection by increasing level by 1. The system selects the items with same rank as the rank of the upper level item for both source and destinations. The searching would continue for both source and destination separately. Those values which have been fetched in this stage would be considered as corresponding SRC and DSTN and the searching would continue as described in the earlier paragraphs.

If no value is found against same rank then the checking would be done for Higher rank and if fail then checking would be performed for lower rank values. Whenever the checking is successful either for source or destination corresponding value would be set as new SRC or new DSTN.

If the system fails in the above process then the system will traverse another upper level to find new SRC and DSTN. This process will continue until new Source and destination is found or the system traverses to the top level of hierarchy.

In the following subsections a function hierarchy of the algorithm, flow-diagram of the algorithm are given and finally the algorithm is given.

3.3.4 Outline of the Algorithm and the Functions Used

In this sub-section we describe the relationship among the different functions of the procedure Lookup. The functionalities of the different functions are also described.

Figure 3.3 depicts the relationship of function hierarchy in *Procedure Lookup*.

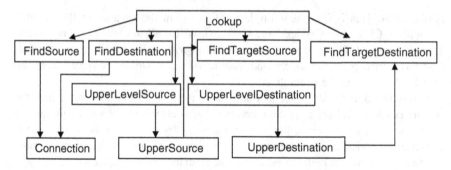

Fig. 3.3 Function hierarchy in Procedure Lookup

3.3.4.1 Purpose of Each Function

(A) FindTargetSource (Src): Fetch the Item with highest rank at target level with respect to given SRC.

(B) FindTargetDestination (DSTN): Fetch the Item with highest rank at target level with respect to given DSTN.

(C) FindSource (SRC, RANK, Select_Source_Item[], DSTN, Select_Destination_Item[]): Fetch the items that are related with SRC and stored in the array Source_Array[].

(D) FindDestination(DSTN, RANK, Select_Destination_Item[], SRC, Select_Source _Item[]): Fetch the items that are related with DSTN and stored in the array Destination_Array[].

(E) Connection (Src, Source_Set_Item, DSTN, Destination_Set_Item): This function is called by the functions FindSource() and FindDestination(). This function finds connection between each Source_Set_Item and DSTN or between each Destination_Set_Item and Source or between each Source_Set_Item and each Destination_Set_Item.

(F) UpperLevelSource(Src): This function is used to check the upper levels of SRC.

(G) UpperSource (Select_Upper_Source_Array_level$_{Level\ Number}$[], Source_Immediate_ Upper_Item_Level$_{Level\ Number}$): This function finds and stores the upper level item of the corresponding item and named it as SRC. Now it calls FindTargetSource (Src).

(H) UpperLevelDestination(dstn): This function is used to check the upper levels of DSTN.

(I) UpperDestination (Select_Upper_Destination_Array_Level$_{Level\ Number}$[], Destination_Immediate_Upper_Item_level$_{Level\ Number}$): It is called by UpperLevelDestination. This function finds and stores the upper level item of the corresponding item and named it as DSTN. Now it calls FindTargetDestination (DSTN).

3.3.5 *Algorithm to Search Information from Concept Hierarchy*

Procedure Lookup (SRC, DSTN)
BEGIN

Step 1:
/* This step is to accept and validate the inputs */
Accept inputs from user;
Validate the input from Search table or the General CHT;
/* The Lookup uses General CHT for all levels if only one CHT is used else the Lookup is done in Search Table*/
IF the input is valid THEN
   IF SRC and DSTN are same THEN
      Display error message "Same Source and Destination";
      Terminate Procedure Lookup;
   ELSE
      Retrieve *Item Id* and *Level Value* of SRC and DSTN from corresponding CHT;
        ENDIF
ELSE
   Display error message "INVALID INPUT";
   Terminate Procedure Lookup;
ENDIF

Step 2:
/* This is to check the level value and to fetch SRC and DSTN node values. */
Step 2.1:
IF Search Table exists THEN /* i.e., multiple CHTs are present */
   IF SRC is both in the Search Table and in target level THEN
      Fetch SOURCE "Item Name", "Item Id", "Rank" from target level CHT;
   ELSE
      Call Function FindTargetSource(SRC);
   End If
   IF DSTN is both in the Search Table and in target level THEN
      Fetch DSTN "Item Name", "Item Id", "Rank" from target level CHT;
   ELSE
       Call Function FindTargetDestination (DSTN)
   ENDIF
ELSE
   Step 2.2:
   /* If only one CHT is there and the General CHT is used in lieu of the Search Table */
   IF the SOURCE "Level Value" = target level in general CHT THEN
      Fetch "Item Name", "Item Id", "Rank" from the General CHT;
   ELSE
      Call Function FindTargetSource(SOURCE);
   ENDIF
   IF DSTN "Level Value" = target level in general CHT THEN
      Fetch "Item Name", "Item Id", "Rank" from the General CHT;
   ELSE
      Call   Function FindTargetDestination(DSTN)

ENDIF
ENDIF. /* this marks the end of Step 2 */

Step 3:
/* This step is to check if there is direct connection between SRC and DSTN */
IF direct connection exists between SRC and DSTN in the connection table of SRC THEN
   Co-Operative Query Language is displayed;
   Terminate Procedure Lookup;
ENDIF

Step 4:
/* This step is to check indirect connection between SRC and DSTN */
Fetch "Rank" of SRC and DSTN from CHT of target Level for both Multiple CHT and
General CHT;
Step 4.1:
IF "Immediate Upper Level Item Id" of any item is same as the SRC THEN
   Retrieve "Item Name", "Item Id", "Rank" from corresponding CHT;
   Keep the items in the array Select_Source_Item[] for SRC;
ENDIF
Step 4.2:
IF "Immediate Upper Level Item Id" of any item is same as the DSTN THEN
   Retrieve "Item Name", "Item Id", "Rank" from corresponding CHT;
   Keep the items in the array Select_Destination_Item[] for DSTN;
ENDIF
Step 4.3:
IF Step 4.1 is executed THEN
   IF any item in Select_Source_Item[] have Rank value *same* as SRC THEN
     FindSource(SRC, RANK, Select_Source_Item[], DSTN, Select _Destination_Item[]);
        /* Here RANK passes the value *same* */
   ENDIF
   IF any item in Select_Source_Item[] have Rank value *greater than* as SRC THEN
     FindSource(SRC, RANK, Select_Source_Item[], DSTN, Select _Destination_Item[]);
     /* Here RANK passes the value *greater than* */
   ENDIF
   IF any item in Select_Source_Item[] have Rank value *less than* as SRC THEN
     FindSource(SRC, RANK, Select_Source_Item[], DSTN, Select _Destination_Item[])
    /* Here RANK passes the value *less than* */
   ENDIF
ENDIF
Step 4.4:
IF Step 4.2 is executed THEN
   IF any item in Select_Source_Item[] have Rank value *same* as DSTN THEN
     FindDestination(DSTN,   RANK,   Select_Destination_Item[],SRC,   Select
  _Source_Item[]);
        /* Here RANK passes the value *same* */
   ENDIF
   IF any item in Select_Source_Item[] have Rank value *greater than* as DSTN THEN

FindDestination(DSTN,   RANK,   Select_Destination_Item[],SRC,   Select
_Source_Item[]);
                    /* Here RANK passes the value *greater than* */
  ENDIF
  IF any item in Select_Source_Item[] have Rank value **less than** as DSTN THEN
      FindDestination(DSTN,   RANK,   Select_Destination_Item[],   SRC,   Select
_Source_Item[]);
                    /* Here RANK passes the value *less than* */
  ENDIF
ENDIF
Step 4.5:
/* In this Step, the Upper level of the Concept Hierarchy are checked to find connection */
IF FindSource(SRC, RANK, Select_Source_Item[], DSTN, Select_Destination _Item[]) is
not called THEN
    Call Function UpperLevelSource(SRC);
    /* this function is for checking the upper levels of Source*/
ENDIF
IF FindDestination(DSTN, RANK, Select_Destination_Item[], SRC, Select_ Source_Item[])
is not called THEN
    Call Function UpperLevelDestination(DSTN)
    /* this function is for checking the upper levels of Destination*/
ENDIF

Step 5:
    Co-Operative Query Language is displayed;
END.

/* FINDING OUT SOURCE_SET_ITEMS */
Function   **FindSource**(SRC,   RANK,   Select_Source_Item[],   DSTN,   Select_
Destination_Item[])
BEGIN
/* RANK could be same, greater or lower*/
/* According to the ranks of the item, the items are stored into an array. The function
Connection() is called to find connection between them */

Step1:
IF item of respective Rank condition present THEN
    Store the items in an array named Source_Array[];
ENDIF
/* Each Item of Source_Array[] is termed as Source_Set_Item hereon */

Step 2:
IF connection is present between SRC and Source_Set_Item (by taking one item at a time
from Source_Array[]) THEN
    Call function Connection (SRC, Source_Set_Item, DSTN, Select_Destination _Item[]) ;
ELSE
    Return
ENDIF
END.

/* FINDING OUT DESTINATION_SET _ITEMS*/
Function   **FindDestination**   (DSTN,   RANK,   Select_Destination_Item[],   SRC,
Select_Source_Item[])
BEGIN
/* RANK could be same, greater or lower*/
/* According to the ranks of the item, the items are stored into an array. The function
Connection() is called to find connection between them */

Step 1:
IF item of respective Rank condition present THEN
    Store those items in an array named Destination_Array[];
ENDIF
/* Each item of Destination_Array[] is termed as Destination_Set_Item hereon */

Step 2:
IF connection is present between DSTN and Destination_Set_item by   taking one item at
a time from Destination_Array[] THEN
    Call function Connection(SRC, Source_Set_Item, DSTN, Select_Destination _Item[]);
ELSE
    Return
ENDIF
END.

/***CHECKING CONNECTIONS BETWEEN THE SOURCE AND DESTINATIONS
THROUGH Source_Set_Item and Destination_Set_Item*/**
Function Connection (SRC, Source_Set_Item, DSTN, Destination_Set_Item)
BEGIN

Step1:
/* This step is done for checking the connection between *Destination and Source_Set_Item*
by taking each *Source_Set_Item* at a time from the *Source_Array[]*/
FOR each Source_Set_Item in the Source_Array[]
    IF direct connection exists between Source_Set_item and DSTN in connection table of
    DSTN THEN
                Return result to step 5 of Procedure Lookup;
        ELSE
                Point at next Source_Set_Item in Source_Array[];
    ENDIF
ENDFOR
Step 1.1:
IF no connection is found after checking to the last element of Source_Array[] THEN

Step 2:
    /* This step is done for checking the connection between Source and Destination_Set_Item
    by taking each Destination_Set_Item at a time from the Destination_Array[]*/
    FOR each Destination_Set_Item in Destination_Array[]
        IF direct connection exists between Destination_Set_item and SRC in the connection
        table of SRC THEN

       Return result to step 5 of Procedure Lookup;
   ELSE
       Point at next Destination_Set_Item in Destination_Array;
   ENDIF
ENDFOR
Step 2.1:
IF no connection is found after checking to the last element of Destination_Array[] THEN
Step 3:
   / This step is done for checking the connection between Source_Set_Item and Destination_Set_Item. This is done by taking each Source_Set_Item at a time from the Source_Array[] and with all the Destination_Set_item in Destination_array*/*
   FOR each Source_Set_Item in the Source_Array[] and for each Destination_Set_Item in the Destination_Array[]
      REPEAT
         IF there exists any direct connection in the connection table of Source_Set_item between Source_Set_Item and Destination_Set_Item THEN
                 Return result to step 5 of Procedure Lookup;
        ELSE
         IF no connection is found after checking all the elements of Source_Array[] and Destination_Array[] THEN
                 Return;
        ENDIF
        Until any connection is found between Source_Set_Item and Destination_Set_Item
      ENDFOR
   ENDIF /* Corresponds to Step 2.1 */
ENDIF /* Corresponds to Step 1.1 */
END.

/* THIS FUNCTION IS WRITTEN TO GET THE ITEM AT THE TARGET LEVEL HAVING HIGHEST RANK WITH RESPECT TO THE SOURCE*/
Function FindTargetSource (SRC)
BEGIN

Step 1:
Decrease SRC level number by 1;
/* The control goes to the immediate lower level of corresponding CHT*/

Step 2:
Repeat the loop of Steps 3 to 6 till the lowest level values are traversed;

Step 3:
Search and select "Item Names" from the corresponding CHT, whose "Immediate Upper Level Item Id" is same as the SRC;

Step 4:
Fetch the "Item Name", "Item Id" , "Rank", "Level Number" into an array named as Fetch_Source_Array_Level$_{Level Number}$[];
   / If this array is for level number 2 then the name of the array would be Fetch_Source_Array _Level$_2$[]*/*

Step 5:
FOR each item of higher level array
   Traverse all the lower level of array one after another;

Step 6:
   Select the array with the lowest level number;
ENDFOR /*End of Step 5 Loop*/
End of Repeat

Step 7:
Choose the "Item Id" with highest "Rank" among them;

Step 8:
FOR the highest Rank Item traverse the lowest level arrays by taking the highest   Rank   Item
at all level till the target level
   Highest_Fetch_Item= Selected item at target level;
ENDFOR
END.

**/* THIS FUNCTION IS WRITTEN TO GET THE ITEM AT THE TARGET LEVEL
HAVING HIGHEST RANK WITH RESPECT TO THE DESTINATION*/**
Function FindTargetDestination (DSTN)
BEGIN

Step1:
Decrease DSTN level number by 1;
/* the control goes to the immediate lower   level of corresponding CHT */

Step 2:
Repeat the loop of Steps 3 to 6 till the lowest level values are traversed

Step 3:
   Search and select "Item Names" from the corresponding CHT whose "Immediate Upper
   Level Item Id" is same as the DSTN;

Step 4:
   Fetch the "Item Name", "Item Id", "Rank" and "Level Number" into an array named as
   Fetch_Destination_Array_Level $_{Level Number}$[];
   /* If this array is for level number 2 then the name of the array would be
   Fetch_Destination_Array_Level$_2$[ ]*/

Step 5:
   FOR each item of higher level array
      Traverse all the lower level of array one after another;

Step 6:
      Select the array with the lowest level number;
   ENDFOR
End of Repeat

Step 7:
Choose the *"Item Id"* with highest "Rank" among them;

Step 8:
FOR the highest Rank Item traverse the lowest level arrays by taking the highest Rank Item at all level till the target level
    Highest_Fetch_Item= Selected item at target level;
ENDFOR
END.

/* CHECKING THE UPPER LEVELS OF SOURCE*/
Function UpperLevelSource(SRC)
BEGIN
/* In this step, the immediate upper level item id is checked for SRC */
Repeat Until Step 3 is evaluated to be true

Step 1:
    Increase the present level number by 1;
    /* The control goes to the immediate upper level CHT */

Step 2:
    Source_immediate_Upper _Item_Level$_{Level\ Number}$= Upper level item with respect to SRC

Step 3:
    IF ImmediateUpperLevelItemId = Source_immediate_Upper _Item_Level$_{Level\ Number}$
    THEN
     Retrieve its Item Name, Item Id, Rank and keep them in an array named as Select_Upper_Source_Array_Level$_{Level\ Number}$[];
       /*If this item is for level number 2 then the name of the item would be Select_Upper_Source_Array_Level$_2$[]*/
     ELSE
     SRC=Source_immediate_Upper _Item_Level$_{Level\ Number}$
    ENDIF
End of Repeat

Step 4:
Retrieve those items from Select_Upper_Source_Array_Level$_{Level\ Number}$[] array which have **same rank** as Source_Immediate_Upper_Item_Level$_{Level\ Number}$;
Call   Function   UpperSource(Select_Upper_Source_Array_level$_{Level\ Number}$[], Source_Immediate_ Upper_Item_Level$_{Level\ Number}$);
Step 4.1
IF Same Rank Item is not present THEN
    Retrieve those items from Select_Upper_Source_Array_Level$_{Level\ Number}$[] array which have **higher rank** than Source_Immediate_Upper_Item_Level$_{Level\ Number}$;
    Call   Function   UpperSource(Select_Upper_Source_Array_level$_{Level\ Number}$[], Source_Immediate_ Upper_Item_Level$_{Level\ Number}$);
ELSE
    Retrieve those items from Select_Upper_Source_Array_Level$_{Level\ Number}$[] array which have **lower rank** than Source_Immediate_Upper_Item_Level$_{Level\ Number}$;

Call   Function   UpperSource(Select_Upper_Source_Array_level$_{Level}$ $_{Number}$[],
Source_Immediate_ Upper_Item_Level$_{Level\ Number}$);
ENDIF
END.

/*FINDING THE PARENT OF EVERY ITEM IN THE CONCEPT HIERARCHY, THAT IS, THE ITEM IN THE IMMEDIATE UPPER LEVEL OF CONCEPT HIERARCHY IS FETCHED HERE*/

Function   UpperSource   (Select_Upper_Source_Array_level$_{Level}$ $_{Number}$[], Source_Immediate_ Upper_Item_Level$_{Level\ Number}$)
BEGIN

Step 1:
Keep the retrieved items of the immediate upper level in an array named as Same_Array$_{Level}$ $_{Number}$[]
/* Each item of Same_Array$_{Level\ Number}$[] is termed as Same_Array$_{Level\ Number}$ hereon */

Step 2:
FOR each Same_Array$_{Level\ Number}$ from the array Same_Array$_{Level\ Number}$[]
    Set Same_Array$_{Level\ Number}$ as SRC;
    SRC = FindTargetSource(SRC);
ENDFOR
END.

/* THE IMMEDIATE UPPER LEVEL ITEM ID IS CHECKED FOR DESTINATION*/
Function UpperLevelDestination(DSTN)
BEGIN
/* The immediate upper level item id is checked for DSTN */
Repeat Until Step 3 is evaluated to be true

Step 1:
    Increase the present level number by 1;
    /* The control goes to the immediate   upper level CHT */

Step 2:
    Destination_Immediate_ Upper_Item_Level$_{Level\ Number}$= Upper level item with respect to DSTN;

Step 3:
    IF *ImmediateUpperLevelItemId* = Destination_Immediate_Upper_Item_Level$_{Level\ Number}$
    THEN
        Retrieve its *Item Name, Item Id, Rank* and keep them in an   array   named   as
        Select_Upper _Destination_Array_Level $_{Level\ Number}$[];
        /* *Here the level number means if this item is for level number 2 then the name is given as   Select_Upper_Destination_Array_Level$_2$[]**/
    ELSE
        DSTN= Destination_Immediate_Upper_Item_Level$_{Level\ Number}$

ENDIF
End of Repeat
Step 4:
Retrieve those items from Select_Upper_Destination_Array_Level$_{Level\ Number}$[] array which have **same rank** as Destination_Immediate_Upper_Item_level$_{Level\ Number}$;
Call   Function   UpperDestination   (Select_Upper_Destination_Array_Level$_{Level\ Number}$[],Destination_Immediate_Upper_Item_level$_{Level\ Number}$)
Step 4.1:
IF Same Rank Item is not present THEN
    Retrieve those items from Select_Upper_Destination_Array_Level$_{Level\ Number}$[] array
    which have **higher rank** than Destination_Immediate_Upper_Item_level$_{Level\ Number}$;
    Call   Function   UpperDestination   (Select_Upper_Destination_Array_Level$_{Level\ Number}$[],Destination_Immediate_Upper_Item_level$_{Level\ Number}$);
ELSE
    Retrieve those items from Select_Upper_Destination_Array_Level$_{Level\ Number}$[] array which
    have **lower rank** than Destination_Immediate_Upper_Item_level$_{Level\ Number}$
    Call   Function   Upper   Destination   (Select_Upper_Destination_Array_Level$_{Level\ Number}$[],Destination_Immediate_Upper_Item_level$_{Level\ Number}$);
ENDIF
END.

/*FINDING THE PARENT OF EVERY ITEM IN THE CONCEPT HIERARCHY, i.e., THE ITEM IN THE IMMEDIATE UPPER LEVEL OF CONCEPT HIERARCHY IS FETCHED HERE*/
Function   UpperDestination   (Select_Upper_Destination_Array_Level$_{Level\ Number}$[],Destination_Immediate_Upper_Item_level$_{Level\ Number}$)

BEGIN

Step 1:
Keep the retrieved items of the immediate upper level in an array named as Same_Array_Level$_{Level\ Number}$[];
/*Each item of Same_Array_Level$_{Level\ Number}$[] is termed as Same_Array$_{Level\ Number}$ here on*/

Step 2:
FOR each Same_Array$_{Level\ Number}$ from the array Same_Array_Level$_{Level\ Number}$[]
    Set Same_Array$_{Level\ Number}$ as DSTN;
    DSTN= FindTargetDestination(DSTN);
ENDFOR
END.

The system performs several searching and checking in the above algorithm. These are performed on the data stored in the different tables in a database or data warehouse system. All of these searching and checking are performed through SQL queries on the tables. Some of these queries generate the result by successfully retrieving the desired data and some of them may not. The set of queries which fetch the desired result successfully are combined together to generate cooperative query answering. The following observation may be noted for the algorithm:

1. Once a solution is found no further level is traversed in concept hierarchy. However from the same level other alternative solutions are given.
2. Higher the level is searched; greater number of alternative solutions could be identified. However, this incurs higher time complexity.
3. If the costs are associated with each activity the algorithm could be slightly modified to compute the total cost of each solution and then could be sorted according to cost.

3.4 Use Case to Illustrate Efficiency of the Algorithm

We have designed and implemented a software-based solution to practically evaluate the effectiveness of our algorithm. The software finds the connections of trains among different stations, cities and states of India. We extracted part of the huge Indian Railway database (4 states out of 29 states in India), enhance with some extra information to express the concept hierarchy and we assigned a rank to each station/city.

Here, concept hierarchy is defined up to 3 levels in the form of Station → City → State. As only one concept hierarchy is defined, General CHT is used in the software development. This software finds the connection from source to destination allowing up to 5 intermediate stations. The Ranks for the items are assigned as follows:

Rank 1: The stations/cities which are the starting stations/cities of the trains which go to other states.
Rank 2: The stations/cities where the trains of other states halt, but not the starting stations/cities.
Rank 3: Other stations/cities.

All the states are assigned with rank 1 as all the stations are connected with other states.

The implementation was carried out by .Net (Visual Studio 2012) as front end and SQL Server 7.0 for the database and the operating system Windows 7. For the experimental result we used an Intel core i3 3.1 GHz processor with the hard disk of 500 GB and RAM of 4 GB.

The stepwise execution of the algorithm is depicted in the following case studies. The execution in all cases was under 10 s.

Case 1: Source is Howrah Station (Station Level) and Destination is Karnataka (State Level) (Howrah is a station in city Kolkata. The rank of Howrah and Kolkata is 1. As Karnataka is a state it has rank 1).

Execution Sequence of algorithm Search:

1. Inputs are validated (Using Step 1 of Lookup)
2. Level is checked from Step 2.2 of the algorithm as the software employs General CHT. As the destination is not in target level FindTargetDestination() is called.
3. In FindTargetDestination(), Fetch_Destination_Array_Level$_2$ is formed for cities of Karnataka. As cities are at level 2 the city information is stored at Fetch_Destination_Array_Level$_2$. This is done at step 4 of the function.
4. As station is the target level, Fetch_Destination_Array_Level$_1$ is formed in next iteration.
5. Again in FindTargetDestination() at Step 8 Highest_Fetch_Item is computed which are the Bangalore City Junction, Yeshvantpur.
6. Now in step 3 of Search the connection table of Source i.e. of Howrah is checked. Connection is found between Howrah and both of the Bangalore City Junction and Yeshvantpur. Hence the connection is shown and algorithm terminate.

Co-Operative Query: Co-operative queries are formed by combining all the queries that are raised in the above steps and executed successfully in the database.

Case 2: Source is Belur (Station Level) and Destination is Mumbai (City Level) (Belur is a small station near Kolkata city. The rank of Belur is 3 and Kolkata is 1. Mumbai is a city with rank 1).

Execution Sequence of algorithm Search:

1. Inputs are validated (Using Step 1 of Lookup)
2. From step 2.2 of Search, "Item Name", "Item Id", "Rank" are fetched from General CHT for Source (Belur) and for Destination (Mumbai) the function FindTargetDestination(DESTINATION) is called.
3. In the function FindTargetDestination(DESTINATION) the array Fetch_Destination_Array_Level$_1$ (As station is in level 1) is formed with the stations of Mumbai. Again from this function Highest_Fetch_Item is computed which are the Mumbai Jn and Mumbai SBC.
4. In step 3 of Search no direct connection is found between Belur and Mumbai Jn or Mumbai SBC.
5. From Step 4.1 and step 4.2 of Search Select_Source_Item[] and Select_Destination_Item[] are formed. In Select_Source_Item[] different items will be added which are connected to Belur and one of them is Howrah which has rank 1. Similarly in Select_Destination_Item[] stations connected with Mumbai Jn and Mumbai SBC are added.
6. Now from step 4.3 of Search FindSource(SOURCE, RANK, Select_Source_Item[], DESTINATION, Select_Destination_Item[]) is called.
7. FindSource calls Connection (Source, Source_Set_Item, DESTINATION, Select_Destination_Item[])

8. In the function Connection () the route is found between Howrah to both Mumbai Jn and Mumbai SBC.
9. This result is returned to Step 5 of Search, which displays the final connection. The route is established as Garia $\rightarrow$ Howrah $\rightarrow$ Mumbai Jn/Mumbai SBC and the algorithm terminate

Co-Operative Query: Co-operative queries are formed by combining all the queries that are raised in the above steps and executed successfully in the database.

3.5 Performance Analysis

Observe that, in order to find out the co-operative query using concept hierarchy, the hierarchical structure needs to be traversed either top-down or bottom-up. Moreover there could be the cases where the structure needs to travel both ways to answer one query. This could be considered as the worst case. Besides, from one node there is exactly one option for going up or going down (as for a particular query one of the concept hierarchy to be followed even if multiple concept hierarchy present).

Hence, traversal would be linear in both ways in the worst case. Thus if we consider total number of representation of a dimension is N, then the graph corresponds to concept hierarchy would consists of N nodes. Therefore if we consider the traversal the concept hierarchy in both ways for single query the worst time complexity would corresponds to $O(N)$.

The performance of this system strictly improves on any other equivalent system which does not incorporate concept hierarchy. There are many different databases or data warehouse based systems which retrieve the data where many attributes are present and some of them also maintain concept hierarchy. However, those systems specially, which are pure database management system don't incorporate the knowledge of concept hierarchy. Hence, user can't make the query at different abstract level of attributes instead they need to make the query in only one form in which the attribute is actually represented in the system. This system overcomes these problems by using the proposed data structure which maintains the concept hierarchy suitably.

Another benefit of this system is that it generates co-operative queries which are often a set of queries that collectively generate the required result set. In comparison to the other system which only generates the result in terms of the single query that is executed, this system outperforms others as it employs the concept of collective query answering by fetching partial results to generate the final result. In this context, we refer to a very popular and useful site of Indian railway system named www.indianrail.gov.in. This shows the availability of trains between two stations in terms of station name or corresponding city name. The queries can be made only based on the city or station name. If the user wants to find the trains among the different districts or states of India, this website can't answer it.

The site www.indianrail.gov.in gives the result if and only if direct connection is present among two cities or stations. We have developed a system based on the proposed algorithm to find the path between source and destination in a manner similar to the Indian railway website. However, this new system is able to find the path among different sources and destinations at different abstract levels (station, city, state). Along with these, if the direct connection is not present, it could still find a connection among the source and destination through intermediate connections. (In the simulated software up to 5 hops are allowed against a journey from source to destination.)

In earlier section 2 case studies have been demonstrated based on the proposed algorithm in the simulated software, which is not possible in www.indianrail.gov. in. In 1st case the destination is in state level and in 2nd case no direct connection is present.

Finally we present a comparative study based on the data collected from real world environment.

3.6 Experimental Verification

3.6.1 Aim of the Experiment

The aim of this experiment is to show the connections among the stations within India. However the input may come in the form State, City or Station. The experiment shows it is capable to work with three abstractions of data.

3.6.2 Technical Environment

The implementation was carried out using .Net (visual Studio 2012) as front end and SQL Server 7.0 to store the data in the database package and the operating system is Windows 7. The hardware specification of the system is as Intel core i3 3.1 GHz with the hard disk of 500 GB and RAM of 4 GB.

3.6.3 Details of Experiment with Result Set

Indian railway database consists of huge amount of operational data. It was not possible to replicate the entire database. Hence we focused on the major cities of India and few States. As we belong to the state of West Bengal the database is populated with details data of West Bengal along with few states like Rajasthan, Karnataka and Kerala.

In order to use the software we surveyed it in 3 classes of our University and told the students about the limited entries of our database compare to Indian railways. They were asked to give the preferred choice of travel places from the given databases. However if there is a repetition of the same choice within a class we asked them to change those data. It is to be noted as majority of our students are from Kolkata (City)/West Bengal (State) higher numbers of entries are from this region.

At first we present three data set and then draw the graph to show how our methodology able to give answer in all the cases, whereas the existing system fails.

The inputs from the users are Source place and Destination Place. However for the convenience of the readers we show the associated concept hierarchy of the given entries. Moreover we display whether the result is successful or not in the Indian railways website.

Data Set-1

Serial no.	Source place	Type of abstraction in concept hierarchy	Destination place	Type of abstraction in concept hierarchy	Result in Indian railway website
1	Kolkata	City	Birbhum	City	No
2	Kolkata	City	Mumbai	City	Yes
3	Chinsurah	Station	Amritsar	City	No
4	Kolkata	City	Tamilnadu	State	No
5	NJP	Station	Kolkata	City	Yes
6	Sealdah	Station	New Delhi	City	Yes
7	Kolkata	City	Barmer	Station	No
8	Kolkata	City	Puri	City	Yes
9	Garia	Station	Mumbai	City	No
10	Kolkata	City	New Delhi	City	Yes
11	Dankuni	Station	New Delhi	City	No
12	Kolkata	City	Bangalore	City	Yes
13	Kolkata	City	Karnataka	State	No
14	Howrah	Station	Namkhana	Station	No
15	NJP	Station	Guwahati	City	Yes
16	Sealdah	Station	Namkhana	Station	Yes
17	NJP	Station	Kalka	Station	No
18	New Delhi	City	Mumbai	City	Yes
19	Kolkata	City	Kalka	Station	Yes
20	New Delhi	City	Rajasthan	State	No

Data Set-2

Serial no.	Source place	Type of abstraction in concept hierarchy	Destination place	Type of abstraction in concept hierarchy	Result in Indian railway website
1	New Delhi	City	Mumbai	City	Yes
2	Jaipur	City	Chandigarh	City	Yes
3	Kolkata	City	NJP	Station	Yes
4	Chennai	City	Trivandrum	City	Yes
5	Sealdah	Station	Burdwan	City	Yes
6	Mumbai	City	NJP	Station	No
7	Guwahati	City	Bhubaneswar	City	Yes
8	Kharagpur	City	Hyderabad	City	Yes
9	New delhi	City	West Bengal	State	No
10	Kolkata	City	Nagpur	City	Yes
11	Jaipur	City	New Delhi	City	Yes
12	Kolkata	City	Asansole	City	Yes
13	Shalimar	Station	NJP	Station	Yes
14	Bangalore	City	Mumbai	City	Yes
15	Kolkata	City	Cuttuck	City	Yes
16	Santragachi	Station	Chennai	City	Yes
17	Bandel	City	Bangalore	City	No
18	Chennai	City	Mysore	City	Yes
19	Kolkata	City	Trivandrum	City	Yes
20	Jaipur	City	Kolkata	City	Yes

Data Set-3

Serial no.	Source place	Type of abstraction in concept hierarchy	Destination place	Type of abstraction in concept hierarchy	Result in Indian railway website
1	Kolkata	City	Puri	City	Yes
2	Belur	Station	Chandigarh	City	No
3	NJP	Station	Kolkata	City	Yes
4	Howrah	Station	Durgapur	City	Yes
5	Santragachi	Station	Trivandrum	City	Yes
6	Kolkata	City	Vizag	City	Yes
7	Mumabi	City	Rajasthan	State	No
8	New delhi	City	Patna	City	Yes
9	Sealdah	Station	Kerala	State	No
10	Gurap	Station	Howrah	Station	Yes
11	Bangalore	City	Chennai	City	Yes
12	Kolkata	City	Jaipur	City	Yes
13	Asansole	City	New Delhi	City	Yes

14	West Bengal	State	Madhya Pradesh	State	No
15	Kolkata	City	Chennai	City	Yes
16	Durgapur	City	NJP	Station	Yes
17	Bhubaneswar	City	Shalimar	Station	Yes
18	Howrah	Station	Garia	Station	No
19	Chennai	City	Santragachi	Station	Yes
20	Kharagpur	City	Mumbai	City	Yes

The above three data sets are executed using the software developed by implementing the proposed algorithms. Based on the result, we plot the graph in Fig. 3.4. This presents a comparative performance based on our proposed algorithm and the Indian railways website. Our software is able to find an answer for all the cases as the source to destination is reachable by 5 intermediate stations and the entries follow the given concept hierarchy (Station → City → State).

The graph in Fig. 3.4 is a performance metric of our software over Indian railways website.

If the entries were given based on Districts of the states or on different Regions of India, the software will not be able to find a solution. However, if the concept hierarchy is modified as Station → City → District → State → Region and the values of District and Region are entered, the system could answer all the cases provided source to destination is reachable by maximum of 5 intermediate stations.

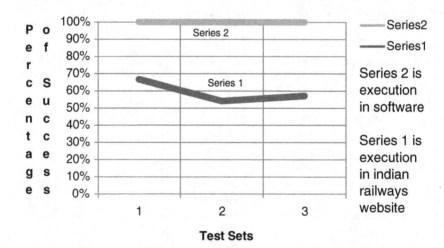

Fig. 3.4 Performance analysis of the software based on proposed algorithm and Indian railways software

3.7 Scalability

While the proposed cooperative query language base solution extracts information using storage hierarchy, this would have a cost as well. The additional traversal in the hierarchy will require more time. This could be prohibitively high for huge data with multiple attributes having concept hierarchy. However, the rank based system incorporated earlier in this chapter enables faster searching for many of these cases. Higher ranking of the source and destination allow better performance. If both of the source and destination has higher ranking, there could be a possibility of direct connection or through minimum intermediaries. Even if only one of the source or destination has high ranking, this would help to search faster as compared to the cases when both are of low ranks. In such cases, majority of the inputs would require traversing the concept hierarchy multiple times and therefore checking a very high number of intermediaries.

3.8 Conclusions

In the experimental result, the software solution that we developed using Concept Hierarchy computes a result for all the cases. It could be a time taking issue if the numbers of intermediary station is high and the volume of data is high. In that case the response time may be an issue of concern. A separate indexing scheme or formation of materialized views may help improving such situation.

References

1. Puthpongsiriporn T (2002) Co-operative query answering for approximate answers with nearness measure in hierarchical structure information systems. Dissertation of Doctor of Philosophy, School of Engineering University of Pittsburgh
2. Blok EH, Choenni S, Blanken MH, Apers GMP (2004) A selectivity model for fragmented relations: applied in information retrieval. IEEE Trans Knowl Data Eng 16(5):635–639
3. Choe G, Nam YK, Goguen J, Wang G (2009) Query generation for retrieving data from distributed semistructured documents using a metadata interface. Elsevier J. Comput Lang Syst Struct 35(4):422–434
4. Siau K, Nah HFF, Cao Q (2011) A meta-analysis comparing relational and semantic models. J Database Manag (JDM) 22(4):57–72
5. Chu WW, Chen Q (1994) A structured approach for cooperative query answering. IEEE Trans Knowl Data Eng 6(5):738–774
6. D'Ulizia A, Ferri F, Grifoni P (2009) Query relaxation in cooperative query processing. In: Methods and supporting technologies for data analysis. Studies in computational intelligence, vol 225, pp 167–185
7. Park H, Rho S, Park J (2011) A link-based ranking algorithm for semantic web resources: a class-oriented approach independent of link direction. J Database Manage (JDM) 22(1):1–25

Chapter 4
Conclusions

The primary motivation of this book is to introduce Hyper-lattice as a new algebraic structure and to establish the context where this handles multi-dimensional data archive in a way which is more flexible and comprehensive as compared to conventional Lattice structures. Thus, Hyper-lattice can be used as a model multi-dimensional data to be stored and accessed for online data analytics.

In the broader perspective, we look to improve the state-of-the-art scenario for efficient storage and access of data warehouse based on OLAP. In order to achieve this goal we have proposed new data model. Algorithms are designed to efficiently traverse the storage structures for data warehouse; query language is used from a distinctly different perspective on existing structure to fetch more information by traversing it.

The areas identified for this book are the lattice of cuboids and concept hierarchy. The proposed *Hyper-lattice* which is not just a contribution towards data warehouse but also may be applied to any domain of science that uses lattice. However, as this book is focused on the application of data warehouse, we discussed different case studies based on business applications to show the effectiveness of this new model for data analytics. Moreover, we have proposed traversal mechanism for efficient storage and retrieval of data in terms of cuboids in Hyper-lattice. Along with this, we also introduced a new type of data warehouse schema named as Hyper-lattice Schema.

The biggest benefit of this new algebraic structure is that even in cases where Lattice suffers for its rigid ad in-flexible structure, Hyper-lattice may still be used to represent associations for multi-dimensional data in a data warehouse. We could finally conclude that improvisation of Hyper-lattice over traditional lattice actually generalizes the lattice by relaxing the conditions.

Just as hierarchy in business data organization is very common and crucial, every dimension which participates in business has multiple abstractions. This is also important to analyze on the basis of all the abstractions. Concept hierarchy

© The Author(s) 2016
S. Sen et al., *Hyper-lattice Algebraic Model for Data Warehousing*,
SpringerBriefs in Applied Sciences and Technology,
DOI 10.1007/978-3-319-28044-8_4

being a hierarchical structure, traversing it both in top-down and bottom-up approaches may help extracting more information. However, this traversal consists of multiple steps single SQL query can't express the queries. Hence we have applied co-operative query to collectively represent more than one query towards fetching more information.

Incorporating concept hierarchy fully on the Hyper-lattice is an extension of this research work. However, the number of cuboids in this case would be excessive and majority of them are useless for business requirements. Thus filtering those cuboids, storing and retrieving the cuboids by defining some sparse based structure could be an interesting scope for future research.

Another interesting work on Hyper-lattice could be managing distributed data warehouse from different sites by aggregating multiple lattices of cuboids those share some common cuboids. Besides, integrating the data warehouse from heterogeneous data sources using Hyper-lattice is also a challenging work.

The study, analysis and improvisation on lattice of cuboids and concept hierarchy for Online Analytical Processing that the book contains would help developing a robust state-of-the-art framework for data analytics.

Index

© The Author(s) 2016
S. Sen et al., *Hyper-lattice Algebraic Model for Data Warehousing*,
SpringerBriefs in Applied Sciences and Technology,
DOI 10.1007/978-3-319-28044-8

Printed in the United States
By Bookmasters